I0073006

De Gruyter Graduate
Tadros • An Introduction to Surfactants

Also of Interest

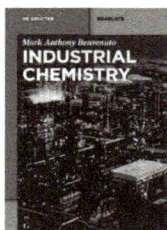

Mark Anthony Benvenuto
Industrial Chemistry, 2013
ISBN 978-3-11-029589-4, e-ISBN 978-3-11-029590-0

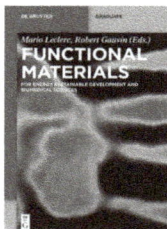

Mario Leclerc, Robert Gauvin (Eds.)
Functional Materials – For Energy, Sustainable Development
and Biomedical Sciences, 2014
ISBN 978-3-11-030781-8, e-ISBN 978-3-11-030782-5

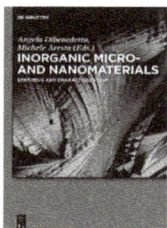

Angela Dibenedetto, Michele Aresta (Eds.)
Inorganic Micro-and Nanomaterials – Synthesis and
Characterization, 2013
ISBN 978-3-11-030666-8, e-ISBN 978-3-11-030687-3,
Set ISBN 978-3-11-030688-0

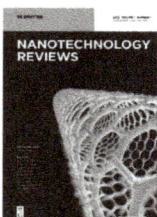

Challa Kumar (Editor-in-Chief)
Nanotechnology Reviews
ISSN 2191-9097

www.degruyter.com

Tharwat F. Tadros

An Introduction to Surfactants

—

DE GRUYTER

Author
Prof. Tharwat F. Tadros
89 Nash Grove Lane
Wokingham RG40 4HE
Berkshire, UK
Email: tharwat@tadros.fsnet.co.uk

ISBN 978-3-11-031212-6
e-ISBN 978-3-11-031213-3

Library of Congress Cataloging-in-Publication Data
A CIP catalog record for this book has been applied for at the Library of Congress.

Bibliographic information published by the Deutsche Nationalbibliothek
The Deutsche Nationalbibliothek lists this publication in the Deutsche Nationalbibliografie;
detailed bibliographic data are available in the Internet at http://dnb.dnb.de.

© 2014 Walter de Gruyter GmbH, Berlin/Boston
Cover image: stasy1980/iStock/Thinkstock
Typesetting: PTP-Berlin, Protago TEX-Produktion GmbH, www.ptp-berlin.de
Printing and binding: Hubert & Co., Göttingen
♾Printed on acid-free paper
Printed in Germany

www.degruyter.com

Preface

Surfactants find application in almost every chemical industry of which the following may be worth mentioning: detergents, paints, dyestuffs, cosmetics, pharmaceuticals, agrochemicals, fibers, plastics, etc. Moreover, surfactants play a major role in the oil industry, for example in enhanced and tertiary oil recovery. They are also occasionally used for environmental protection, e.g. in oil slick dispersants. Therefore, a fundamental understanding of the physical chemistry of surface active agents, their unusual properties and their phase behavior is essential for most industrial chemists. In addition, an understanding of the basic phenomena involved in the application of surfactants, such as in the preparation of emulsions and suspensions and their subsequent stabilization, in nanoemulsions, in microemulsions, in wetting spreading and adhesion, etc., is of vital importance in arriving at the right composition and control of the system involved This book has been written as an introduction to graduate chemists, industrial formulators who are using surfactants in the above applications. As far as possible, the book is written in a simple way without too much detail in order to enable the reader to make a start with the basic physical phenomena involved in such a vast field. For more detail, the reader can refer to more comprehensive books edited or written by the author (such as "Applied Surfactants", Wiley, 2005; "Encyclopedia of Colloid and Interface Science", Springer, 2013).

This book starts with a general introduction (Chapter 1) to introduce the subject and explain the book structure. Chapter 2 gives a general classification of surfactants based on the nature of the head group (anionic, cationic, zwitterionic and nonionic), a description of some specialized molecules such as fluorocarbon and silicone surfactants (referred to as superwetters), sugar-based surfactants, naturally occurring surfactants and polymeric surfactants. Chapter 3 deals with the unusual properties of surfactant solutions and the process of micellization. The different self-assembly structures that are produced in surfactant solutions are described in terms of their structures and phase behavior. Chapter 4 describes the process of surfactant adsorption at the air/liquid (A/L), liquid/liquid (L/L) and solid/liquid (S/L) interfaces. The experimental techniques that can be applied for measuring surfactant adsorption at various interfaces are briefly described. Chapter 5 describes the use of surfactants as emulsifiers with particular attention to the methods that can be applied for selection of emulsifiers for a given oil used in the emulsion. The role of surfactants in stabilizing the emulsion against flocculation, Ostwald ripening and coalescence is also described at a fundamental level. Chapter 6 describes the use of surfactants as dispersants for suspensions. The process of dispersion of powders in liquids is described in terms of wetting, dispersion and stabilization of the resulting suspension against flocculation. Chapter 7 deals with surfactants in foam formation and stabilization. The theories of foam stabilization and the role of surfactants are described. Application of surfactants in formulation of nanoemulsions (with size range 20–200 nm) is described in

Chapter 8. The various methods that can be applied for preparation of nanoemulsions are described. Chapter 9 describes microemulsions and the origin of their thermodynamic stability. The various methods that can be applied for formulation of microemulsions are described. The use of surfactants as wetting agents will be described in Chapter 10. Particular attention is given to the fundamentals of wetting and spreading with particular reference to surfactants that can be used as wetting agents. The final Chapter 11 deals with applications of surfactants in selected industries: Cosmetics – Pharmaceuticals – Agrochemicals – Paints and Coatings – Detergents.

This book can be useful for teaching undergraduate and graduate students. It is also valuable for industrial chemists who are involved in formulation of disperse systems, where surfactants are essential ingredients in such formulations.

Berkshire, 2014
Tharwat F. Tadros

Contents

1 General introduction

Surface active agents (usually referred to as surfactants) are amphiphilic or amphipathic molecules consisting of a non-polar hydrophobic portion, usually a straight or branched hydrocarbon or fluorocarbon chain containing 8–18 carbon atoms, which is attached to a polar or ionic portion (hydrophilic). The term amphiphilic originates from the Greek word "amphi", meaning "both" and the term relates to the fact that all surfactant molecules consist of at least two parts, one which is soluble in a specific fluid, e.g. water (the hydrophilic part) and one which is insoluble in water (the hydrophobic part). The hydrophilic portion can be nonionic, ionic or zwitterionic, accompanied by counter ions in the last two cases. The hydrocarbon chain interacts weakly with the water molecules in an aqueous environment, whereas the polar or ionic head group interacts strongly with water molecules via dipole or ion-dipole interactions. It is this strong interaction with the water molecules which renders the surfactant soluble in water. However, the water molecules avoid contact with the hydrophobic chain and their cooperative action of dispersion and hydrogen bonding tends to squeeze the hydrocarbon chain out of the water by accumulation at interfaces and association in solution to form aggregate units referred to as micelles. In the latter case, the surfactant hydrophobic groups are directed towards the interior of the aggregate and the polar head groups are directed towards the solvent. These micelles are in dynamic equilibrium and the rate of exchange between a surfactant molecule and the micelle may vary by orders of magnitude, depending on the structure of the surfactant molecule. The balance between hydrophilic and hydrophobic parts of the molecule (referred to as the hydrophilic-lipophilic balance, HLB) gives these systems their special properties such as adsorption at interfaces and formation of self-assembly structures.

Surfactants have the property of adsorbing onto the surfaces or interfaces of the system and of altering the surface or interfacial free energy of those surfaces or interfaces. The driving force for surfactant adsorption is the lowering of the free energy of the phase boundary. The interfacial free energy per unit area is the amount of work required to expand the interface. This interfacial free energy, referred to as surface or interfacial tension, γ, is given in mJm^{-2} or mNm^{-1}. Adsorption of surfactant molecules at the interface lowers the surface tension γ_{AW} (at the air/liquid interface) or interfacial tension γ_{OW} (at the oil/water interface) and the higher the surfactant adsorption (i.e. the more dense the layer is) the larger the reduction in γ. Surfactants also adsorb at the solid/liquid interface and this causes a reduction in the solid/liquid interfacial tension, γ_{SL}. The degree of surfactant adsorption at the interface depends on surfactant structure and the nature of the two phases that meet the interface [1–4].

When studying surfactants one should consider two main phenomena:
1. Interfacial effects that relates to the adsorption and orientation of the molecules at various interfaces. This requires accurate measurements of the adsorption and orientation of the surfactant ions or molecules.
2. Colloid stability that relates to the effect of surfactants on stabilization of various disperse systems, e.g. emulsions, suspensions, foams, nanoemulsions and microemulsions. It should be mentioned that this subdivision is only for convenience since colloid and interface science are one and the same subject of study. All colloid stability phenomena are related to the interfacial phenomena.

Surfactants find application in almost every chemical industry of which the following may be worth mentioning: detergents, paints, dyestuffs, cosmetics, pharmaceuticals, agrochemicals, fibers, plastics, etc. Moreover, surfactants play a major role in the oil industry, for example in enhanced and tertiary oil recovery. In the latter case surfactant micellar systems and microemulsions are used to recover oil from micro-pores that has been entrapped as a result of capillary forces. They are also occasionally used for environmental protection, e.g. in oil slick dispersants. The spilled oil from tankers and oil wells is emulsified using surfactants and the resulting emulsion is separated and then the system is demulsified to recover the oil. Therefore, a fundamental understanding of the physical chemistry of surface active agents, their unusual properties and their phase behavior is essential for most industrial chemists. In addition, an understanding of the basic phenomena involved in the application of surfactants, such as in the preparation of emulsions and suspensions and their subsequent stabilization, in nanoemulsions, in microemulsions, in wetting spreading and adhesion, etc., is of vital importance in arriving at the right composition and control of the system involved [1, 2]. This is particularly the case with many formulations in the chemical industry mentioned above.

It should be stated that commercially produced surfactants are not pure chemicals, and within each chemical type there can be tremendous variation. This can be understood, since surfactants are prepared from various feed stocks, namely petrochemicals, natural vegetable oils and natural animal fats. It is important to realize that in every case the hydrophobic group exists as a mixture of chains of different lengths. The same applies to the polar head group, for example in the case of polyethylene oxide (the major component of nonionic surfactants) which consists of a distribution of ethylene oxide units. Hence, products that may be given the same generic name could vary a great deal in their properties and the formulation chemist should bear this in mind when choosing a surfactant from a particular manufacturer. It is advisable to obtain as much information as possible from the manufacturer, such as the distribution of alkyl chain length, distribution of the polyethylene oxide chain and also the properties of the surfactant chosen such as its suitability for the job, its batch to batch variation, its toxicity, etc. The manufacturer usually has more information on the surfactant than that printed in the data sheet, and in most cases such information is given on request.

This book gives an introduction to surfactants, their solution properties, adsorption at various interfaces and their applications in various disperse systems. Chapter 2 gives a general classification of surfactants based on the nature of the head group (anionic, cationic, zwitterionic and nonionic). Description of some specialized molecules, such as fluorocarbon and silicone surfactants (referred to as superwetters), and sugar-based surfactants is also given. Naturally occurring surfactants that are used in the food industry and pharmaceuticals are also described. A section will be devoted to polymeric surfactants. The latter are particularly important for stabilization of disperse systems. Chapter 3 deals with the unusual properties of surfactant solutions that show abrupt changes at a particular concentration that is related to the formation of aggregate units referred to as micelles. This concentration that is referred to as the critical micelle concentration (CMC) depends on the structure and nature of the surfactant molecule. The different self-assembly structures that are produced in surfactant solutions are described in terms of their structures and phase behavior. Chapter 4 describes the process of surfactant adsorption at the air/liquid (A/L), liquid/liquid (L/L) and solid/liquid (S/L) interfaces. A thermodynamic treatment of the process of surfactant adsorption is given. This treatment can be applied for the reversible adsorption of the surfactant molecules whereby an equilibrium is established when the rate of adsorption becomes equal to the rate of desorption. Such thermodynamic treatment cannot be applied for the adsorption of polymeric surfactants since in this case the adsorption process is not reversible. For that case statistical thermodynamic treatment of the process of adsorption can be applied. The experimental techniques that can be applied for measuring surfactant adsorption at various interfaces are briefly described. Understanding the process of surfactant adsorption at various interfaces is very important in their application. For example, the process of wetting and spreading on various interfaces is determined by the adsorption of surfactant molecules at the A/L and S/L interfaces. Adsorption at the L/L interfaces determines the process of emulsification and emulsion stability. The same applies for nanoemulsions and microemulsions. Chapter 5 describes the use of surfactants as emulsifiers; particular attention is given to the methods that can be applied for selection of emulsifiers for a given oil used in the emulsion. The role of surfactants in stabilizing the emulsion against flocculation, Ostwald ripening and coalescence is also described at a fundamental level. Chapter 6 describes the use of surfactants as dispersants for suspensions. The process of dispersion of powders in liquids is described in terms of wetting, dispersion and stabilization of the resulting suspension against flocculation. The process of Ostwald ripening (crystal growth) and the role of surfactants is described at a fundamental level. Chapter 7 deals with surfactants in foam formation and stabilization. The theories of foam stabilization and the role of surfactants are described. Application of surfactants in formulation of nanoemulsions is described in Chapter 8. Nanoemulsions are a special class of emulsions with droplet size in the range 20–200 nm. Their main advantages in formulation are described. The various methods that can be applied for preparation of nanoemulsions are described. Chapter 9 describes microemulsions and the origin of their thermodynamic stability. The various methods that can be applied for formulation of microemulsions are described. The use of surfactants as wetting agents will be described in Chapter 10. Particular attention is given to the fundamentals of wetting and spreading with particular reference to surfactants that can be used as wetting agents. The final Chapter 11 will deal with application of surfactants in various industries: Cosmetics – Pharmaceuticals – Agrochemicals – Paints and Coatings – Detergency – Oil Recovery.

It should be mentioned that this book is written for graduate students and scientists who are beginners in the subject. As far as possible, the subject is dealt with at a fundamental level without too much detail. For more comprehensive understanding of the subject of surfactants, the reader can refer to other texts that are given in the reference list.

References

[1]   Th. F. Tadros (ed.), *Surfactants*, Academic Press, London, 1984.
[2]   M. R. Porter, *Handbook of Surfactants*, Chapman and Hall, Blackie, USA, 1994.
[3]   K. Holmberg, B. Jonsson, B. Kronberg and B. Lindman, *Surfactants and Polymers in Solution*, second edition, John Wiley and Sons, Ltd., Chichester, UK, 2003.
[4]   Th. Tadros, *Applied Surfactants*, Wiley-VCH, Germany, 2005.
[5]   M. J. Rosen and J.T. Kunjappu, *Surfactants and Interfacial Phenomena*, John Wiley and Sons, USA, 2012.
[6]   Th. Tadros (ed.), *Encyclopedia of Colloid and Interface Science*, Springer, Germany, 2013.

2 General classification of surfactants

A simple classification of surfactants based on the nature of the hydrophilic group is commonly used. Four main classes may be distinguished, namely anionic, cationic, amphoteric. and nonionic [1, 2]. A useful technical reference is McCutcheon [3], which is produced annually to update the list of available surfactants. A recent text by van Os et al. [4] listing the physicochemical properties of selected anionic, cationic and nonionic surfactants has been published by Elsevier. Another useful text is the Handbook of Surfactants by Porter [5]. It should also be mentioned that a fifth class of surfactants, usually referred to as polymeric surfactants, has been used for many years for preparation of emulsions and suspensions and their stabilization.

2.1 Anionic surfactants

These are the most widely used class of surfactants in industrial applications [5–7]. This is due to their relatively low cost of manufacture and they are practically used in every type of detergent. For optimum detergency the hydrophobic chain is a linear alkyl group with a chain length in the region of 12–16 C atoms and the polar head group should be at the end of the chain. Linear chains are preferred since they are more effective and more degradable than the branched chains. The most commonly used hydrophilic groups are carboxylates, sulfates, sulfonates and phosphates. A general formula may be ascribed to anionic surfactants as follows:

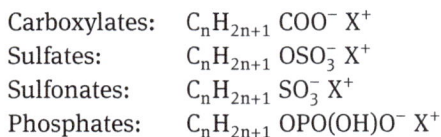

Carboxylates: $C_nH_{2n+1} COO^- X^+$
Sulfates: $C_nH_{2n+1} OSO_3^- X^+$
Sulfonates: $C_nH_{2n+1} SO_3^- X^+$
Phosphates: $C_nH_{2n+1} OPO(OH)O^- X^+$

with n being the range 8–16 atoms and the counterion X^+ is usually Na^+.

Several other anionic surfactants are commercially available such as sulfosuccinates, isethionates (esters of isothionic acid with the general formula $RCOOCH_2-CH_2-SO_3Na$) and taurates (derivatives of methyl taurine with the general formula $RCON(R')CH_2-CH_2-SO_3Na$), sarchosinates (with the general formula $RCON(R')COONa$) and these are sometimes used for special applications. Below a brief description of the above anionic classes is given with some of their applications.

2.1.1 Carboxylates

These are perhaps the earliest known surfactants, since they constitute the earliest soaps, e.g. sodium or potassium stearate, $C_{17}H_{35}COONa$, sodium myristate, $C_{14}H_{29}COONa$. The alkyl group may contain unsaturated portions, e.g. sodium oleate, which contains one double bond in the C_{17} alkyl chain. Most commercial soaps will be a mixture of fatty acids obtained from tallow, coconut oil, palm oil, etc. They are simply prepared by saponification of the triglycerides of oils and fats The main attraction of these simple soaps is their low cost, their ready biodegradability and low toxicity. Their main disadvantage is their ready precipitation in water containing bivalent ions such as Ca^{2+} and Mg^{2+}. To avoid their precipitation in hard water, the carboxylates are modified by introducing some hydrophilic chains, e.g. ethoxy carboxylates with the general structure $RO(CH_2CH_2O)_nCH_2COO^-$, ester carboxylates containing hydroxyl or multi-COOH groups, sarcosinates which contain an amide group with the general structure $RCON(R')COO^-$. The addition of the ethoxylated groups results in increased water solubility and enhanced chemical stability (no hydrolysis). The modified ether carboxylates are also more compatible with electrolytes. They are also compatible with other nonionic, amphoteric and sometimes even cationic surfactants. The ester carboxylates are very soluble in water, but they suffer from the problem of hydrolysis. The sarcosinates are not very soluble in acid or neutral solutions but they are quite soluble in alkaline media. They are compatible with other anionics, nonionics and cationics. The phosphate esters have very interesting properties being intermediate between ethoxylated nonionics and sulfated derivatives. They have good compatibility with inorganic builders and they can be good emulsifiers. A specific salt of a fatty acid is lithium 12-hydroxystearic acid that forms the major constituent of greases.

2.1.2 Sulfates

These are the largest and most important class of synthetic surfactants, which are produced by reaction of an alcohol with sulfuric acid, i.e. they are esters of sulfuric acid. In practice sulfuric acid is seldom used and chlorosulfonic or sulfur dioxide/air mixtures are the most common methods of sulfating the alcohol. However, due to their chemical instability (hydrolyzing to the alcohol, particularly in acid solutions), they are now overtaken by the sulfonates which are chemically stable. The properties of sulfate surfactants depend on the nature of the alkyl chain and the sulfate group. The alkali metal salts show good solubility in water, but they tend to be affected by the presence of electrolytes. The most common sulfate surfactant is sodium dodecyl sulfate (abbreviated as SDS and sometimes referred to as sodium lauryl sulfate) which is extensively used both for fundamental studies as well as in many applications in industry. At room temperature (~ 25 °C) this surfactant is quite soluble and 30 % aqueous solutions are fairly fluid (low viscosity). However, below 25 °C, the surfactant may separate out as a soft paste as the temperature falls below its Krafft point (the temperature above which the surfactant shows a rapid increase in solubility with further increase of temperature). The latter depends on the distribution of chain lengths in the alkyl chain: the wider the distribution the lower the Krafft temperature. Thus, by controlling this distribution one may achieve a Krafft temperature of ~ 10 °C. As the surfactant concentration is increased to 30–40 % (depending on the distribution of chain length in the alkyl group), the viscosity of the solution increases very rapidly and may produce a gel but then falls at about 60–70 % to give a pourable liquid, after which it increases again to a gel. The concentration at which the minimum occurs varies according to the alcohol sulfate used, and also the presence of impurities such as unsaturated alcohol. The viscosity of the aqueous solutions can be reduced by addition of short chain alcohols and glycols. The critical micelle concentration (cmc) of SDS (the concentration above which the properties of the solution show abrupt changes) is 8×10^{-3} mol dm^{-3} (0.24 %). The alkyl sulfates give good foaming properties with an optimum at C_{12}–C_{14}. As with the carboxylates, the sulfate surfactants are also chemically modified to change their properties. The most common modification is to introduce some ethylene oxide units in the chain, usually referred to as alcohol ether sulfates. These are made by sulfation of ethoxylated alcohols. For example, sodium dodecyl 3-mole ether sulfate which is essentially dodecyl alcohol reacted with 3 moles EO, then sulfated and neutralized by NaOH. The presence of PEO confers improved solubility when compared with the straight alcohol sulfates. In addition, the surfactant becomes more compatible with electrolytes in aqueous solution. The ether sulfates are also more chemically stable than the alcohol sulfates. The cmc of the ether sulfates is also lower than the corresponding surfactant without the EO units. The viscosity behavior of aqueous solutions is similar to alcohol sulfates, giving gels in the range 30–60 %. The ether sulfates show a pronounced salt effect, with significant increase of the viscosity of a dilute solution on addition of electrolytes such as NaCl. The ether sulfates are commonly used in hand dishwashing and in shampoos in combination with amphoteric surfactants.

2.1.3 Sulfonates

With sulfonates, the sulfur atom is directly attached to the carbon atom of the alkyl group and this gives the molecule stability against hydrolysis, when compared with the sulfates (whereby the sulfur atom is indirectly linked to the carbon of the hydrophobe via an oxygen atom). The alkyl aryl sulfonates are the most common type of these surfactants (for example sodium alkyl benzene sulfonate) and these are usually prepared by reaction of sulfuric acid with alkyl aryl hydrocarbons, e.g. dodecyl benzene. A special class of sulfonate surfactants are the naphthalene and alkyl naphthalene sulfonates which are commonly used as dispersants. As with the sulfates, some chemical modification is used by introducing ethylene oxide units, e.g. sodium nonyl phenol 2-mole ethoxylate ethane sulfonate $C_9H_{19}C_6H_4(OCH_2CH_2)_2SO_3^-Na^+$. The paraffin sulfonates are produced by sulfo-oxidation of normal linear paraffins with sulfur dioxide and oxygen and catalyzed with ultraviolet or gamma radiation. The resulting alkane sulfonic acid is neutralized with NaOH. These surfactants have excellent water solubility and biodegradability. They are also compatible with many aqueous ions. The linear alkyl benzene sulfonates (LABS) are manufactured from alkyl benzene and the alkyl chain length can vary from C_8 to C_{15} and their properties are mainly influenced by the average molecular weight and the spread of carbon number of the alkyl side chain. The cmc of sodium dodecyl benzene sulfonate is 5×10^{-3} mol dm^{-3} (0.18 %). The main disadvantage of LABS is their effect on the skin and hence they cannot be used in personal care formulations.

Another class of sulfonates is the α-olefin sulfonates which are prepared by reacting linear α-olefin with sulfur trioxide, typically yielding a mixture of alkene sulfonates (60–70 %), 3- and 4-hydroxyalkane sulfonates (~ 30 %) and some disulfonates and other species. The two main α-olefin fractions used as starting material are C_{12}–C_{16} and C_{16}–C_{18}. Fatty acid and ester sulfonates are produced by sulfonation of unsaturated fatty acids or esters. A good example is sulfonated oleic acid,

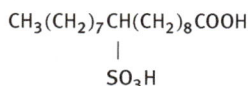

$$CH_3(CH_2)_7CH(CH_2)_8COOH$$
$$|$$
$$SO_3H$$

A special class of sulfonates are sulfosuccinates which are esters of sulfosuccinic acid,

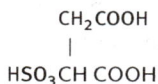

$$CH_2COOH$$
$$|$$
$$HSO_3CH\,COOH$$

Both mono- and diesters are produced. A widely used diester in many formulations is sodium di(2-ethylhexyl)sulfosuccinate (that is sold commercially under the trade name Aerosol OT). The cmc of the diesters is very low, in the region of 0.06 % for C_6–C_8 sodium salts and they give a minimum surface tension of 26 mNm^{-1} for the C_8 diester. Thus these molecules are excellent wetting agents. The diesters are soluble both in water and in many organic solvents. They are particularly useful for preparation of water-in-oil (W/O) microemulsions.

2.1.4 Isethionates

These are esters of isethionic acid $HOCH_2CH_2SO_3H$. They are prepared by reaction of acid chloride (of the fatty acid) with sodium isethionate. The sodium salt of C_{12-14} are soluble at high temperature (70 °C) but they have very low solubility (0.01 %) at 25 °C. They are compatible with aqueous ions and hence they can reduce the formation of scum in hard water. They are stable at pH 6–8 but they undergo hydrolysis outside this range. They also have good foaming properties.

2.1.5 Taurates

These are derivatives of methyl taurine $CH_2-NH-CH_2-CH_2-SO_3$. The latter is made by reaction of sodium isethionate with methyl amine. The taurates are prepared by reaction of fatty acid chloride with methyl taurine. Unlike the isethionates, the taurates are not sensitive to low pH. They have good foaming properties and they are good wetting agents.

2.1.6 Phosphate-containing anionic surfactants

Both alkyl phosphates and alkyl ether phosphates are made by treating the fatty alcohol or alcohol ethoxylates with a phosphorylating agent, usually phosphorous pentoxide, P_4O_{10}. The reaction yields a mixture of mono- and diesters of phosphoric acid. The ratio of the two esters is determined by the ratio of the reactants and the amount of water present in the reaction mixture. The physicochemical properties of the alkyl phosphate surfactants depend on the ratio of the esters. They have properties intermediate between ethoxylated nonionics (see below) and the sulfated derivatives. They have good compatibility with inorganic builders and good emulsifying properties. Phosphate surfactants are used in the metal working industry due to their anticorrosive properties.

2.2 Cationic surfactants

The most common cationic surfactants are the quaternary ammonium compounds [8, 9] with the general formula $R'R''R'''R''''N^+X^-$, where X^- is usually a chloride ion and R represents alkyl groups. These quaternaries are made by reacting an appropriate tertiary amine with an organic halide or organic sulfate. A common class of cationics is the alkyl trimethyl ammonium chloride, where R contains 8–18 C atoms, e.g. dodecyl trimethyl ammonium chloride, $C_{12}H_{25}(CH_3)_3NCl$. Another widely used cationic surfactant class is that containing two long chain alkyl groups, i.e. dialkyl dimethyl ammonium chloride, with the alkyl groups having a chain length of 8–18 C atoms. These dialkyl surfactants are less soluble in water than the monoalkyl quaternary compounds, but they are commonly used in detergents as fabric softeners. A widely used cationic surfactant is alkyl dimethyl benzyl ammonium chloride (sometimes referred to as benzalkonium chloride and widely used as bactericide), having

the structure

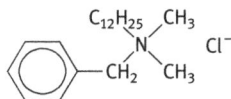

$$
\underset{\text{(benzene ring)}}{\bigcirc}\!\!-\!\!CH_2-\overset{\displaystyle C_{12}H_{25}}{\underset{\displaystyle CH_3}{\overset{|}{\underset{|}{N}}}}\!\!\overset{CH_3}{\underset{}{}}\quad Cl^-
$$

Imidazolines can also form quaternaries, the most common product being the ditallow derivative quaternized with dimethyl sulfate,

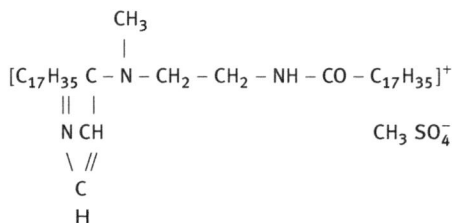

$$
\left[C_{17}H_{35}\,\underset{\substack{|| \; | \\ N \; CH \\ \;\; \backslash \; // \\ \;\;\; C \\ \;\;\; H}}{C-\overset{\displaystyle CH_3}{\overset{|}{N}}-CH_2-CH_2-NH-CO-C_{17}H_{35}} \right]^{+} \qquad CH_3\,SO_4^-
$$

Cationic surfactants can also be modified by incorporating polyethylene oxide chains, e.g. dodecyl methyl polyethylene oxide ammonium chloride having the structure,

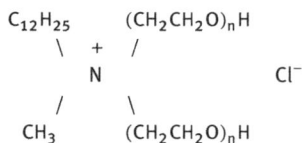

$$
\begin{array}{c}
C_{12}H_{25}\qquad\quad (CH_2CH_2O)_nH \\
\;\;\;\backslash \;\; + \;\; / \\
\;\;\;\;\;\;\;\;N \qquad\qquad\qquad Cl^- \\
\;\;\;/ \qquad\quad \backslash \\
CH_3 \qquad\quad (CH_2CH_2O)_nH
\end{array}
$$

Cationic surfactants are generally water soluble when there is only one long alkyl group. When there are two or more long chain hydrophobes, the product becomes dispersible in water and soluble in organic solvents. They are generally compatible with most inorganic ions and hard water, but they are incompatible with metasilicates and highly condensed phosphates. They are also incompatible with protein-like materials. Cationics are generally stable to pH changes, both acid and alkaline. They are incompatible with most anionic surfactants, but they are compatible with nonionics. These cationic surfactants are insoluble in hydrocarbon oils. In contrast, cationics with two or more long alkyl chains are soluble in hydrocarbon solvents, but they become only dispersible in water (sometimes forming bilayer vesicle type structures). They are generally chemically stable and can tolerate electrolytes. The cmc of cationic surfactants is close to that of anionics with the same alkyl chain length. For example, the cmc of benzalkonium chloride is 0.17 %. The prime use of cationic surfactants is their tendency to adsorb at negatively charged surfaces, e.g. anticorrosive agents for steel, flotation collectors for mineral ores, dispersants for inorganic pigments, antistatic agents for plastics, antistatic agents and fabric softeners, hair conditioners, anticaking agent for fertilizers and as bactericides.

2.3 Amphoteric (zwitterionic) surfactants

These are surfactants containing both cationic and anionic groups [10]. The most common amphoterics are the N-alkyl betaines which are derivatives of trimethyl glycine $(CH_3)_3NCH_2COOH$ (that was described as betaine). An example of a betaine surfactant is lauryl amido propyl dimethyl betaine $C_{12}H_{25}CON(CH_3)_2CH_2COOH$. These alkyl betaines are sometimes described as alkyl dimethyl glycinates. The main characteristic of amphoteric surfactants is their dependence on the pH of the solution in which they are dissolved. In acid pH solutions, the molecule acquires a positive charge and it behaves like a cationic, whereas in alkaline pH solutions they become negatively charged and behave like an anionic. A specific pH can be defined at which both ionic groups show equal ionization (the isoelectric point of the molecule). This can be described by the following scheme,

$$N^+ \ldots COOH \quad \leftrightarrow \quad N^+ \ldots COO^- \quad \leftrightarrow \quad NH \ldots COO^-$$

acid pH < 3   isoelectric   pH > 6 alkaline

Amphoteric surfactants are sometimes referred to as zwitterionic molecules. They are soluble in water, but the solubility shows a minimum at the isoelectric point. Amphoterics show excellent compatibility with other surfactants, forming mixed micelles. They are chemically stable both in acids and alkalis. The surface activity of amphoterics varies widely and it depends on the distance between the charged groups and they show a maximum in surface activity at the isoelectric point.

Another class of amphoteric are the N-alkyl amino propionates having the structure $R-NHCH_2CH_2COOH$. The NH group is reactive and can react with another acid molecule (e.g. acrylic) to form an amino dipropoionate $R-N(CH_2CH_2COOH)_2$. Alkyl imidazoline-based product can also be produced by reacting alkyl imidozoline with a chloro acid. However, the imidazoline ring breaks down during the formation of the amphoteric.

The change in charge with pH of amphoteric surfactants affects their properties, such as wetting, detergency, foaming, etc. At the isoelectric point, the properties of amphoterics resemble those of nonionics very closely. Below and above the i.e.p, the properties shift towards those of cationic and anionic surfactants respectively. Zwitterionic surfactants have excellent dermatological properties. They also exhibit low eye irritation and they are frequently used in shampoos and other personal care products (cosmetics). Due to their mild characteristics, i.e. low eye and skin irritation, amphoterics are widely used in shampoos. They also provide antistatic properties to hair, good conditioning and foam booster.

2.4 Nonionic surfactants

The most common nonionic surfactants are those based on ethylene oxide, referred to as ethoxylated surfactants [11–13]. Several classes can be distinguished: alcohol ethoxylates, alkyl phenol ethoxylates, fatty acid ethoxylates, monoalkaolamide ethoxylates, sorbitan ester ethoxylates, fatty amine ethoxylates and ethylene oxide-propylene oxide copolymers (sometimes referred to as polymeric surfactants). Another important class of nonionics are the multihydroxy products such as glycol esters, glycerol (and polyglycerol) esters, glucosides (and polyglucosides) and sucrose esters. Amine oxides and sulfinyl surfactants represent nonionics with a small head group.

2.4.1 Alcohol ethoxylates

These are generally produced by ethoxylation of a fatty chain alcohol such as dodecanol. Several generic names are given to this class of surfactants such as ethoxylated fatty alcohols, alkyl polyoxyethylene glycol, monoalkyl polyethylene oxide glycol ethers, etc. A typical example is dodecyl hexaoxyethylene glycol monoether with the chemical formula $C_{12}H_{25}(OCH_2CH_2O)_6OH$ (sometimes abbreviated as $C_{12}E_6$). In practice, the starting alcohol will have a distribution of alkyl chain lengths and the resulting ethoxylate will have a distribution of ethylene oxide (EO) chain length. Thus the numbers listed in the literature refer to average numbers.

The cmc of nonionic surfactants is about two orders of magnitude lower than the corresponding anionics with the same alkyl chain length. At a given alkyl chain length, the cmc decreases with decrease in the number of EO units. The solubility of the alcohol ethoxylates depend both on the alkyl chain length and the number of ethylene oxide units in the molecule. Molecules with an average alkyl chain length of 12 C atoms and containing more than 5 EO units are usually soluble in water at room temperature. However, as the temperature of the solution is gradually raised, the solution becomes cloudy (as a result of dehydration of the PEO chain and the change in the conformation of the PEO chain) and the temperature at which this occurs is referred to as the cloud point (CP) of the surfactant. At a given alkyl chain length, the CP increases with an increase in the EO chain of the molecule. The CP changes with change of concentration of the surfactant solution and the trade literature usually quotes the CP of a 1 % solution. The CP is also affected by the presence of electrolyte in the aqueous solution. Most electrolytes lower the CP of a nonionic surfactant solution. Nonionics tend to have maximum surface activity near to the cloud point. The CP of most nonionics increases markedly on the addition of small quantities of anionic surfactants. The surface tension of alcohol ethoxylate solutions decreases with an increase in its concentration, until it reaches its cmc, after which it remains constant with further increase in its concentration. The minimum surface tension reached at and above the cmc decreases with decrease in the number of EO units of the chain (at a given alkyl chain). The viscosity of a nonionic surfactant solution increases gradually with an increase in its concentration, but at a critical concentration (which depends on the alkyl and EO chain length) the viscosity shows a rapid increase and ultimately a gel-like structure appears. This results from the formation of liquid crystalline structure of the hexagonal type. In many cases, the viscosity reaches a maximum after which it shows a decrease due to the formation of other structures (e.g. lamellar phases) (see below).

2.4.2 Alkyl phenol ethoxylates

These are prepared by reaction of ethylene oxide with the appropriate alkyl phenol. The most common surfactants of this type are those based on nonyl phenol. These surfactants are cheap to produce, but they suffer from the problem of biodegradability and potential toxicity (the by-product of degradation is nonyl phenol which has considerable toxicity for fish and mammals). In spite of these problems, nonyl phenol ethoxylates are still used in many industrial properties, due to their advantageous properties, such as their solubility both in aqueous and nonaqueous media, their good emulsification and dispersion properties, etc.

2.4.3 Fatty acid ethoxylates

These are produced by reaction of ethylene oxide with a fatty acid or a polyglycol and they have the general formula $RCOO-(CH_2CH_2O)_nH$. When a polyglycol is used, a mixture of mono- and diesters $RCOO-(CH_2CH_2O)_n-OCOR)$ is produced. These surfactants are generally soluble in water provided there are enough EO units and the alkyl chain length of the acid is not too long. The monoesters are much more soluble in water than the diesters. In the latter case, a longer EO chain is required to render the molecule soluble. The surfactants are compatible with aqueous ions, provided there is not much unreacted acid. However, these surfactants undergo hydrolysis in highly alkaline solutions.

2.4.4 Sorbitan esters and their ethoxylated derivatives (Spans and Tweens)

The fatty acid esters of sorbitan (generally referred to as Spans, an Atlas commercial trade name) and their ethoxylated derivatives (generally referred to as Tweens) are perhaps one of the most commonly used nonionics. They were first commercialized by Atlas in the USA, which has been purchased by ICI. The sorbitan esters are produced by reaction of sorbitol with a fatty acid at a high temperature ($> 200\,°C$). The sorbitol dehydrates to 1,4-sorbitan and then esterification takes place. If one mole of fatty acid is reacted with one mole of sorbitol, one obtains a monoester (some diester is also produced as a by-product). Thus, sorbitan monoester has the following general formula:

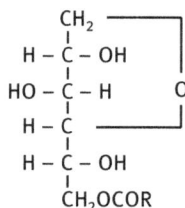

$$
\begin{array}{l}
CH_2 \underline{\hspace{2cm}} \\
H - C - OH \\
HO - C - H \qquad\qquad O \\
H - C \underline{\hspace{1.5cm}} \\
H - C - OH \\
CH_2OCOR
\end{array}
$$

The free OH groups in the molecule can be esterified, producing di- and tri-esters. Several products are available depending on the nature of the alkyl group of the acid and whether the product is a mono-, di- or triester. Some examples are given below,

Sorbitan monolaurate	– Span 20
Sorbitan monopalmitate	– Span 40
Sorbitan monostearate	– Span 60
Sorbitan monooleate	– Span 80
Sorbitan tristearate	– Span 65
Sorbitan trioleate	– Span 85

The ethoxylated derivatives of Spans (Tweens) are produced by reaction of ethylene oxide on any hydroxyl group remaining on the sorbitan ester group. Alternatively, the sorbitol is first ethoxylated and then esterified. However, the final product has different surfactant properties to the Tweens. Some examples of Tween surfactants are given below,

Polyoxyethylene (20) sorbitan monolaurate]	– Tween 20
Polyoxyethylene (20) sorbitan monopalmitate]	– Tween 40
Polyoxyethylene (20) sorbitan monostearate]	– Tween 60
Polyoxyethylene (20) sorbitan monooleate]	– Tween 80
Polyoxyethylene (20) sorbitan tristearate]	– Tween 65
Polyoxyethylene (20) sorbitan trioleate]	– Tween 85

The sorbitan esters are insoluble in water, but soluble in most organic solvents (low HLB number surfactants). The ethoxylated products are generally soluble in number and they have relatively high HLB numbers. One of the main advantages of the sorbitan esters and their ethoxylated derivatives is their approval as food additives. They are also widely used in cosmetics and some pharmaceutical preparations.

2.4.5 Ethoxylated fats and oils

A number of natural fats and oils have been ethoxylated, e.g. lanolin (wool fat) and castor oil ethoxylates. These products are useful for applications in pharmaceutical products, e.g. as solubilizers.

2.4.6 Amine ethoxylates

These are prepared by addition of ethylene oxide to primary or secondary fatty amines. With primary amines both hydrogen atoms on the amine group react with ethylene oxide and therefore the resulting surfactant has the structure,

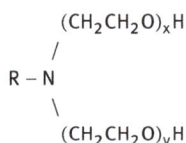

$$
R-N
\begin{cases}
(CH_2CH_2O)_xH \\
(CH_2CH_2O)_yH
\end{cases}
$$

The above surfactants acquire a cationic character if the EO units are small in number and if the pH is low. However, at high EO levels and neutral pH they behave very similarly to nonionics. At low EO content, the surfactants are not soluble in water, but become soluble in an acid solution. At high pH, the amine ethoxylates are water soluble provided the alkyl chain length of the compound is not long (usually a C_{12} chain is adequate for reasonable solubility at sufficient EO content.

2.4.7 Amine oxides

These are prepared by oxidizing a tertiary nitrogen group with aqueous hydrogen peroxide at temperatures in the region 60–80 °C. Several examples can be quoted: N-alkyl amidopropyl-dimethyl amine oxide, N-alkyl bis(2-hydroxyethyl) amine oxide and N-alkyl dimethyl amine oxide. They have the general formula

$$
CocoCONHCH_2CH_2CH_2N
\begin{array}{c}
CH_3 \\
| \\
\rightarrow O \\
| \\
CH_3
\end{array}
$$
  Alkyl amidopropyl-dimethyl amine oxide

CH$_2$CH$_2$OH
|
Coco N → O   Coco bis (2-hydroxyethyl) amine oxide
|
CH$_2$CH$_2$OH

                CH$_3$
                |
CH$_{12}$H$_{25}$ N → O   Lauryl dimethyl amine oxide
                |
                CH$_3$

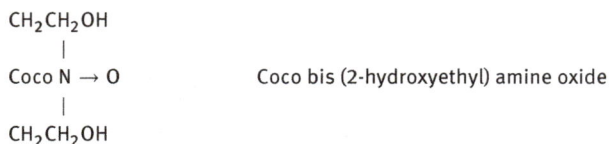

In acid solutions, the amino group is protonated and acts as a cationic surfactant. In neutral or alkaline solution the amine oxides are essentially nonionic in character. Alkyl dimethyl amine oxides are water soluble up to C$_{16}$ alkyl chain. Above pH 9, amine oxides are compatible with most anionics. At pH 6.5 and below some anionics tend to interact and form precipitates. In combination with anionics, amine oxides can be foam boosters (e.g. in shampoos).

2.5 Speciality surfactants

2.5.1 Fluorocarbon and silicone surfactants

These surfactants can lower the surface tension of water to values below 20 mNm^{-1}. Most surfactants described above lower the surface tension of water to values above 20 mNm^{-1}, typically in the region of 25–27 mNm^{-1}. The fluorocarbon and silicone surfactants are sometimes referred to as superwetters as they cause enhanced wetting and spreading of their aqueous solution. However, they are much more expensive than conventional surfactants and they are only applied for specific applications whereby the low surface tension is a desirable property. Fluorocarbon surfactants have been prepared with various structures consisting of perfluoroalkyl chains and anionic, cationic, amphoteric and polyethylene oxide polar groups. These surfactants have good thermal and chemical stability and they are excellent wetting agents for low energy surfaces. Silicone surfactants, sometimes referred to as organosilicones are those with polydimethylsilixane backbone. The silicone surfactants are prepared by incorporation of a water-soluble or hydrophilic group into a siloxane backbone. The latter can also be modified by incorporation of a paraffinic hydrophobic chain at the end or along the polysiloxane backbone. The most common hydrophilic groups are EO/PO and the structures produced are rather complex and most manufacturers of silicone surfactants do not reveal the exact structure. The mechanism by which these molecules lower the surface tension of water to low values is far from being well understood. The surfactants are widely applied as spreading agents on many hydrophobic surfaces. Incorporating organophilic groups into the backbone of the polydimethyl siloxane backbone can give products that exhibit surface active properties in organic solvents.

2.5.2 Gemini surfactants

A gemini surfactant is a dimeric molecule consisting of two hydrophobic tails and two head groups linked together with a short spacer [14]. This is illustrated below for a molecule containing two cationic head groups (separated by two methylene groups) with two alkyl chains,

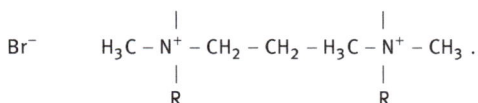

$$\text{Br}^- \qquad H_3C - N^+ - CH_2 - CH_2 - H_3C - N^+ - CH_3 \ .$$

with R substituents on each N^+.

These surfactants show several interesting physicochemical properties, such as very high efficiency in lowering the surface tension and very low cmc. For example, the cmc of a conventional cationic dodecyltrimethylammonium bromide is 16 mM, whereas that of the corresponding gemini surfactant, having a two carbon linkage between the head groups, is 0.9 mM. In addition, the surface tension reached at and above the cmc is lower for gemini surfactants when compared to that of the corresponding conventional one. Gemini surfactants are also more effective in lowering the dynamic surface tension (time required for reaching the equilibrium value is shorter). These effects are due to the better packing of the gemini surfactant molecules at the air/water interface.

2.5.3 Surfactants derived from mono- and polysaccharides

Several surfactants were synthesized starting from mono- or oligosaccharides by reaction with the multifunctional hydroxyl groups: alkyl glucosides, alkyl polyglucosides [15], sugar fatty acid esters and sucrose esters [16], etc. The technical problem is one of joining a hydrophobic group to the multihydroxyl structure. Several surfactants were made, e.g. esterification of sucrose with fatty acids or fatty glycerides to produce sucrose esters having the following structure:

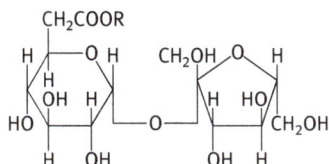

The most interesting sugar surfactants are the alkyl polyglucosides (APG) which are synthesized using a two stage transacetalization process [15]. In the first stage, the carbohydrate reacts with a short chain alcohol, e.g. butanol or propylene glycol. In the second stage, the short chain alkyl glucoside is transacetalized with a relatively long chain alcohol (C_{12-14}–OH) to form the required alkyl polyglucoside. This process is applied if oligo- and polyglucoses (e.g. starch, syrups with a low dextrose equivalent, DE) are used. In a simplified transacetalization process, syrups with high glucose content (DE > 96 %) or solid glucose types can react with short-chain alcohols under normal pressure. The scheme for alkyl polygluciside synthesis is shown below. Commercial alkyl polyglucosides (APG) are complex mixtures of species varying in the degree of polymerization (DP, usually in the range 1.1–3) and in the length of the alkyl chain. When the latter is shorter than C_{14}, the product is water soluble. The cmc values of APG's are comparable to nonionic surfactants and they decrease with increasing alkyl chain length.

APG surfactants have good solubility in water and they have high cloud points (> 100 °C). They are stable in neutral and alkaline solutions but are unstable in strong acid solutions. APG surfactants can tolerate high electrolyte concentrations and they are compatible with most types of surfactants. They are used in personal care products for cleansing formulations as well as for skin care and hair products. They are also used in hard-surface cleaners and laundry detergents. Several applications in agrochemical formulations can be mentioned such as wetting agents and penetrating agents for the active ingredient.

2.5.4 Naturally occurring surfactants

Several naturally occurring amphipathic molecules (in the body) exist, such as bile salts, phospholipids, cholesterol, which play an important role in various biological processes. Their interactions with other solutes, such as drug molecules, and with membranes are also very important. Bile salts are synthesized in the liver and they consist of alicyclic compounds possessing hydroxyl and carboxyl groups. As an illustration, the structure of cholic acid is given below:

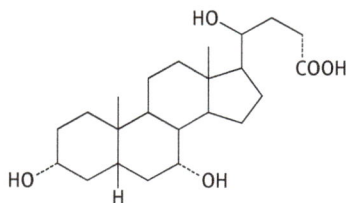

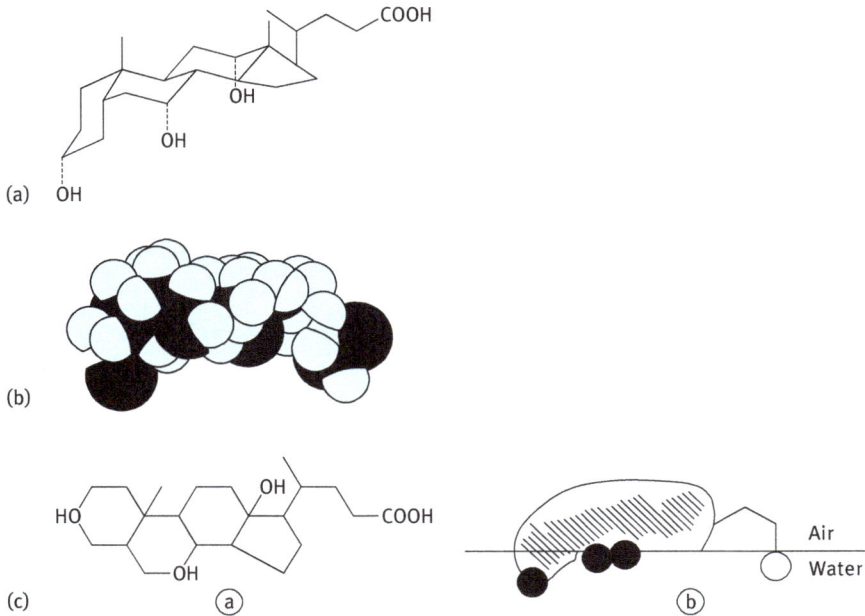

Fig. 2.1. (a) Structural formula of cholic acid showing the cis position of the A ring; (b) Courtauld space filling model of cholic acid; (c) orientation of cholic acid molecules at the air-water interface (hydroxyl groups represented by filled circles and carboxylic acid groups by open circles).

It is the positioning of the hydrophilic groups in relation to the hydrophobic steroidal nucleus that gives the bile salts their surface activity and determines the ability to aggregate. Fig. 2.1 shows the possible orientation of cholic acid at the air-water interface, the hydrophilic groups being oriented towards the aqueous phase [17, 18]. The steroid portion of the molecule is shaped like a "saucer" as the A ring is cis with respect to the B ring. Small [19] suggested that small or primary aggregates with up to 10 monomers form above the cmc by hydrophobic interactions between the non-polar side of the monomers. These primary aggregates form larger units by hydrogen bonding between the primary micelles. This is schematically illustrated in Fig. 2.2. Oakefull and Fisher [18–20] stressed the role of hydrogen bonding rather than hydrophobic bonding in the association of bile salts. However, Zana [21] regarded the association as a continuous process with hydrophobic interaction as the main driving force.

The cmc of bile salts is strongly influenced by its structure; the trihydroxy cholanic acids have higher cmc than the less hydrophilic dihydroxy derivatives. As expected, the pH of solutions of these carboxylic acid salts has an influence on micelle formation. At sufficiently low pH, bile acids which are sparingly soluble will be precipitated from solution, initially being incorporated or solubilized in the existing micelles. The pH at which precipitation occurs, on saturation of the micellar system, is generally

Longitudinal
section

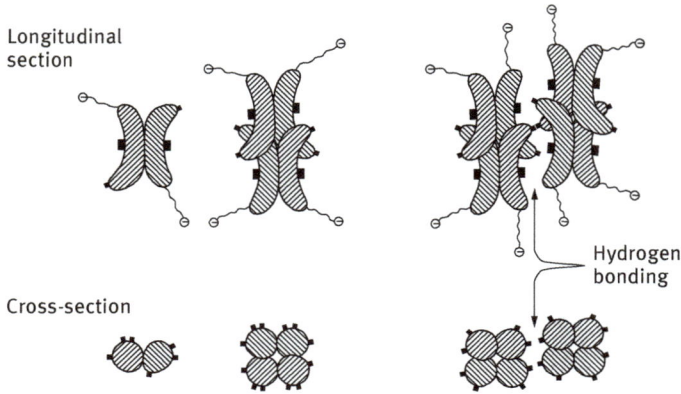

Hydrogen
bonding

Cross-section

Fig. 2.2. Schematic representation of the structure of bile acid salt micelles.

about one pH unit higher than the pK_a of the bile acid. Bile salts play important roles in physiological functions and drug absorption. It is generally agreed that bile salts aid fat absorption. Mixed micelles of bile salts, fatty acids and monogylycerides can act as vehicles for fat transport. However, the role of bile salts in drug transport is not well understood. Several suggestions have been made to explain the role of bile salts in drug transport, such as facilitation of transport from liver to bile by direct effect on canicular membranes, stimulation of micelle formation inside the liver cells, binding of drug anions to micelles, etc. The enhanced absorption of medicinals on administration with deoxycholic acid may be due to reduction in interfacial tension or micelle formation. The administration of quinine and other alkaloids in combination with bile salts has been claimed to enhance their parasiticidal action. Quinine, taken orally, is considered to be absorbed mainly from the intestine and a considerable amount of bile salts is required to maintain a colloidal dispersion of quinine. Bile salts may also influence drug absorption either by affecting membrane permeability or by altering normal gastric emptying rates. For example, sodium taurcholate increases the absorption of sulfaguanidine from the stomach, jejunum and ileum. This is due to an increase in membrane permeability induced by calcium depletion and interference with the bonding between phospholipids in the membrane.

Another important naturally occurring class of surfactants which are widely found in biological membranes are the lipids, such as phosphatidylcholine (lecithin), lysolecithin, phosphatidylethanolamine and phospahitidyl inositol. The structure of these lipids is given in Fig. 2.3. These lipids are also used as emulsifiers for intravenous fat emulsions, anesthetic emulsions as well as for production of liposomes or vesicles for drug delivery. The lipids form coarse turbid dispersions of large aggregates (liposomes) which on ultrasonic irradiation form smaller units or vesicles. The liposomes are smectic mesophases of phospholipids organized into bilayers which assume a multilamellar or unilamellar structure. The multilamellar species are heterogeneous aggregates, most commonly prepared by dispersal of a thin film of phospholipid (alone or with cholesterol) into water. Sonication of the multilamellar units can produce the unilamellar liposomes, sometimes referred to as vesicles. The net charge of liposomes can be varied by incorporation of a long chain amine, such as stearyl amine (to give a positively charged vesicle) or dicetyl phosphate (giving negatively charged species). Both lipid-soluble and water-soluble drugs can be entrapped in liposomes. The liposoluble drugs are solubilized in the hydrocarbon interiors of the lipid bilayers, whereas the water-soluble drugs are intercalated in the aqueous layers. The use of liposomes as drug carriers has been reviewed by Fendler and Romero [26], to which the reader should refer for details. Liposomes, like micelles, may provide a special medium for reactions to occur between the molecules intercalated in the lipid bilayers or between the molecules entrapped in the vesicle and free solute molecules.

Phospholipids play an important role in lung functions. The surface active material to be found in the alveolar lining of the lung is a mixture of phospholipids, neutral lipids and proteins. The lowering of surface tension by the lung surfactant system and the surface elasticity of the surface layers assists alveolar expansion and contraction. Deficiency of lung surfactants in newborns leads to a respiratory distress syndrome and this led to the suggestion that instillation of phospholipid surfactants could cure the problem.

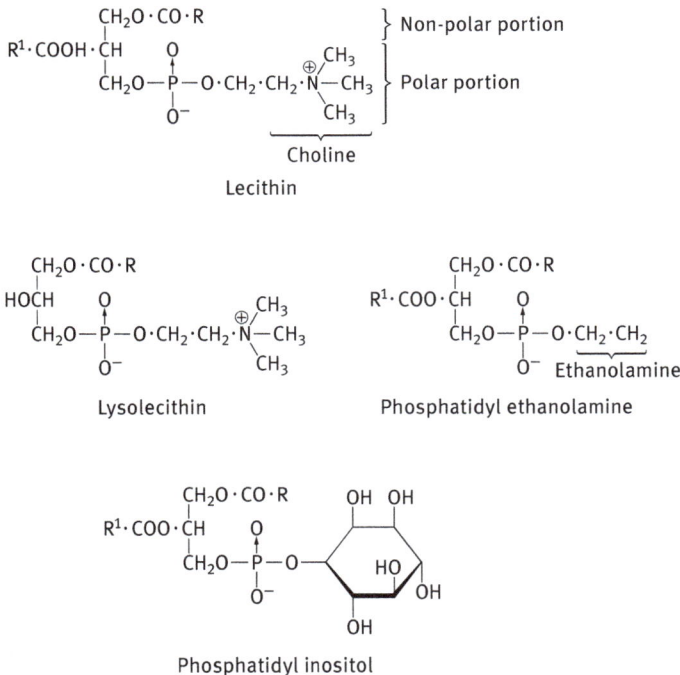

Fig. 2.3. Structure of lipids.

2.5.5 Biosurfactants

In recent years there has been a great deal of concern regarding the use of conventional surfactants in cosmetic and personal care applications. This is due to the environmental and health concerns when using many of the currently used synthetic surfactants. These concerns can be alleviated by the use of biosurfactants that are produced from natural raw materials that possess good biodegradability, low toxicity and the desirable functional performance such as good emulsification and dispersion, high physical stability and high performance on application. The biosurfactants are produced using catalysts in the form of living microorganisms and enzymes [22–24]. Microbial biosurfactants are structurally diverse and complex and are produced in a biosynthetic root catalyzed by enzymes. Several classes are available: glycolipids, e.g. rhamnolipids, sophorolipids; lipopeptides and lipoproteins, e.g. surfactin, polymyxins, gramicidins; phospholipids and fatty acids; and complex combinations of biopolymers, e.g. emulsan, liposan. The enzymatic synthesis of surfactants are essentially chemical reactions in which an enzyme replaces a conventional chemical catalyst. The surfactants produced by a single enzyme are simpler in structure when compared to those produced by microorganisms. Enzymes are able to catalyze a wide range of reactions and they are unique in their specificity and selectivity. The selectivity is recognized at three levels: chemo-, regio- and enantioselectivity. Thus reactions that cannot be achieved by classical organic synthesis (that may require several steps) can be facilitated by biocatalysis. The high selectivity results in the production of fewer by-products. Development of solvent-free processes is also possible using enzymes and hence the products are safer and environmentally friendly. Biocatalysis has been used for synthesis of surfactants with different hydrophilic groups that are attached to the hydrophobic chain via ester, amide or glycisidic bonds. Typical examples are the mono- and diacyl-glycerols. The technical production uses a natural fat or oil (triglyceride) as the starting material. Enzymatic methods have the advantage of high selectivity. For example, it is possible to make 2-monoacyl glycerols by selective removal of the fatty acids in the 1- and 3-positions using a regiospecific lipase as a catalyst in a suitable reaction medium. Enzymes can be used to modify glycerolphospholipids by removal of one of the fatty acids to produce lysophospholipids (a good emulsifying agent). These transformations can be achieved using phospholipase A2 that removes the fatty acid in the sn-2 position.

Lipases have been used to synthesize a wide range of surfactant esters from fatty acids and carbohydrates. The fatty acid can be used either in the free form or as an ester. To make it possible, one needs to find a solvent that efficiently dissolves both fatty acid and carbohydrate. The carbohydrate is made more hydrophobic by converting it to an acetal leaving one OH group for esterification. The acetal groups are removed by acid catalysis. This is illustrated in Scheme 1 in Fig. 2.4. Alternatively the carbohydrate is made more hydrophobic by converting it to a butyl glycoside using glucosisade which is then further reacted with fatty acid using lipase as is illustrated

Scheme 1

Scheme 2

Fig. 2.4. Schemes for synthesis of an alkyl glycoside by a two-step enzymatic root.

in Scheme 2 in Fig. 2.4. Most studies of enzymatic sugar ester synthesis have been focused on the esterification of monosaccharides, since the problem with poor substrate miscibility increases significantly with increasing the size of the carbohydrate. However, by careful choice of reaction conditions it has been possible to acylate several di- and trisaccharides. Solvent mixtures of 2-methyl-2-butanol and dimethylsulfoxide have been used in combination with vinyl esters (C8–C18) to make successful acylation of maltose, maltotriose and leucrose. Ionic liquids are promising nonaqueous solvents for the dissolution of carbohydrates and they have been used in several studies on enzymatic sugar ester synthesis. Protease-catalyzed synthesis of sugar esters has been reported. 6-O-butyl-D-glucose was prepared using subtilisin (a protease from Bacillus subtilis) as catalyst and trichloroethyl butyrate as acyl donor in anhydrous dimethyl formamide. Oligosaccharides as long as maltoheptase were also acylated under these conditions. Bacillus protease has been used to synthesize sucrose laurate from sucrose and vinyl laurate in dimethylformamide. Lipases can accept a wide variety of nucleophiles for the deacylation of acyl enzyme synthesis. By proper choice of reaction conditions it was possible to acylate a wide range of carbohydrates. The main products are obtained by esterification of the primary hydroxyl group. In carbohydrates having more than one primary OH group, the enzyme can selectively acylate one of them. This is the case with maltose. With fructose, mixtures are obtained. Most carbohydrate esters have been prepared using monosaccharides as substrates.

Fatty acid derivatives with an amide bond possess useful properties for surfactants, e.g. enhancing foaming properties of cleaning and personal care products, stabilizing foam and enhancing detergency. The amide bond increases the hydrophilicity of the fatty acid and therefore the surfactant becomes chemically and physically very stable under alkaline conditions. Synthesis of amides can be done by proteases, but these enzymes are very specific for certain amino acids and are more sensitive to organic solvents. Lipases have been used for synthesis of peptides, fatty amides and N-acylamino acids and the acylation of alkanolamines.

Another important class of biosurfactants are the amino-sugar derivatives. Glycamide surfactants are nonionic, biodegradable and in which the hydrophilic moiety (an amino-sugar derivative) is linked to the fatty acid by an amide bond, e.g. glucamides and lactobionamides. A conventional method for preparing sugar fatty amide surfactants includes the Schotton–Baumann reaction between an amine and a fatty acid chloride, where the chloride salt produced has to be removed. The regio- and enantioselectivity of enzymes provides a convenient method of acylation of sugars and sugar amines. Chemoselective acylation of a secondary amine, N-methyl glucamine with fatty acid is possible using lipases (from Novozym).

Alkanolamides are important fatty acid derivates for a wide range of applications, e.g. personal care and hard-surface cleaning. They are characterized by their skin tolerance, good biodegradability and low toxicity. Alkanolamides are produced by condensation of fatty acids or fatty acid esters or triglyceride with alkanolamine, e.g. monoethanolamine or diethanolamine, using high temperature or a metal oxide catalyst. Alkanolamine synthesis can also be achieved using lipase. Alkanolamines are susceptible to acylation both at the amine and hydroxyl group. The main product using lipase (Novozym 435) is the amide.

Amino acid/peptide-lipid conjugates are an interesting class of surfactants with good surface activity, excellent emulsifying agents, antimicrobial activity, low toxicity and high biodegradability. They are attractive for application in personal care products. The large variety of amino acid/peptide structures combined with fatty acids of varying structure and carbon chain length can produce surfactants with wide structural diversity and different physicochemical and biological properties. Depending on the free functional groups on the amino acids, anionic, nonionic, amphoteric and cationic surfactants can be produced. Different forms of amino acid surfactants have been synthesized using enzymes.

As mentioned before, carbohydrate ester surfactants suffer from the disadvantage of chemical instability at neutral or alkaline pH (due to the instability of the ester bond). This instability problem can be overcome by using alkyl glycosides which are based on similar building blocks. Chemical synthesis is useful for the production of mixtures of alkyl glycosides as discussed before. When pure isomers are needed, enzymatic synthesis offers an attractive alternative. Two main classes of enzymes can be used for coupling the carbohydrate part to a hydrophobic alcohol, namely glycosyl transferases and glycosyl hydrolases. An alternative route to alkyl glycosides is to use transglycsylation reaction with an activated carbohydrate substrate

2.5.6 Polymeric surfactants

The simplest type of a polymeric surfactant is a homopolymer that is formed from the same repeating units, such as poly(ethylene oxide) or poly(vinyl pyrrolidone). These homopolymers have little surface activity at the o/w interface, since the homopolymer segments (ethylene oxide or vinylpyrrolidone) are highly water soluble and have little affinity to the interface. However, such homopolymers may adsorb significantly at the S/L interface. Even if the adsorption energy per monomer segment to the surface is small (fraction of kT, where k is the Boltzmann constant and T is the absolute temperature), the total adsorption energy per molecule may be sufficient to overcome the unfavorable entropy loss of the molecule at the S/L interface. Clearly, homopolymers are not the most suitable emulsifiers or dispersants. A small variant is to use polymers that contain specific groups that have high affinity to the surface. This is exemplified by partially hydrolyzed poly(vinyl acetate) (PVAc), technically referred to as poly(vinyl alcohol) (PVA). The polymer is prepared by partial hydrolysis of PVAc, leaving some residual vinyl acetate groups. Most commercially available PVA molecules contain 4–12 % acetate groups. These acetate groups, which are hydrophobic, give the molecule its amphipathic character. On a hydrophobic surface such as polystyrene, the polymer adsorbs with preferential attachment of the acetate groups on the surface, leaving the more hydrophilic vinyl alcohol segments dangling in the aqueous medium. These partially hydrolyzed PVA molecules also exhibit surface activity at the o/w interface [25]. The most convenient polymeric surfactants are those of the block and graft copolymer type. A block copolymer is a linear arrangement of blocks of variable monomer composition. The nomenclature for a diblock is poly-A-block-poly-B and for a triblock is poly-A-block-poly-B-poly-A. One of the most widely used triblock polymeric surfactants are the "Pluronics" (BASF, Germany), which consists of two poly-A blocks of poly(ethylene oxide) (PEO) and one block of poly(propylene oxide) (PPO). Several chain lengths of PEO and PPO are available. Two types may be distinguished: those prepared by reaction of polyoxypropylene glycol (difunctional) with EO or mixed EO/PO, giving block copolymers with the structure,

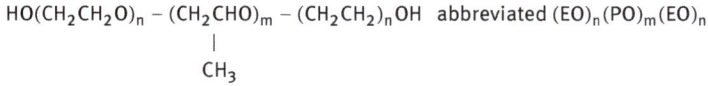

$$HO(CH_2CH_2O)_n - (CH_2CHO)_m - (CH_2CH_2)_nOH \quad \text{abbreviated } (EO)_n(PO)_m(EO)_n$$
$$|$$
$$CH_3$$

Various molecules are available, where n and m are varied systematically. The second type of EO/PO copolymers are prepared by reaction of polyethylene glycol (difunctional) with PO or mixed EO/PO. These will have the structure $(PO)_n(EO)_m(PO)_n$ and they are referred to as reverse Pluronics. These polymeric triblocks can be applied as emulsifiers or dispersants, whereby the assumption is made that the hydrophobic PPO chain resides at the hydrophobic surface, leaving the two PEO chains dangling in aqueous solution and hence providing steric repulsion. Trifunctional products are also available where the starting material is glycerol. These have the structure,

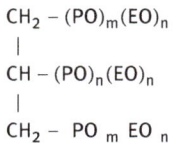

$$CH_2 - (PO)_m(EO)_n$$
$$|$$
$$CH - (PO)_n(EO)_n$$
$$|$$
$$CH_2 - PO_m EO_n$$

Tetrafunctional products are available where the starting material is ethylene diamine. These have the structures,

$(EO)_n$ \qquad\qquad $(EO)_n$

\qquad\quad \\ \qquad\qquad\qquad /

NCH_2CH_2N

\qquad\quad / \qquad\qquad\qquad \\

$(EO)_n$ \qquad\qquad $(EO)_n$

$(EO)_n(PO)_m$ \qquad\qquad $(PO)_m(EO)_n$

\qquad\qquad \\ \qquad\qquad /

NCH_2CH_2N

\qquad\qquad / \qquad\qquad \\

$(EO)_n(PO)_m$ \qquad\qquad $(PO)_m(EO)_n$

Although these block polymeric surfactants have been widely used in various applications in emulsions and suspensions, some doubt has arisen on how effective these can be. It is generally accepted that the PPO chain is not sufficiently hydrophobic to provide a strong "anchor" to a hydrophobic surface or to an oil droplet. Indeed, the reason for the surface activity of the PEO–PPO–PEO triblock copolymers at the o/w interface may stem from a process of "rejection" anchoring of the PPO chain since it is not soluble both in oil and water. Several other di- and triblock copolymers have been synthesized, although these are of limited commercial availability. Typical examples are diblocks of polystyrene-block-polyvinyl alcohol, triblocks of poly(methyl methacrylate)-block poly(ethylene oxide)-block poly(methyl methacrylate), diblocks of polystyrene block-polyethylene oxide and triblocks of polyethylene oxide-block polystyrene-polyethylene oxide [25]. An alternative (and perhaps more efficient) polymeric surfactant is the amphipathic graft copolymer consisting of a polymeric backbone B (polystyrene or polymethyl methacrylate) and several A chains ("teeth") such as polyethylene oxide [25] This graft copolymer is sometimes referred to as a "comb"

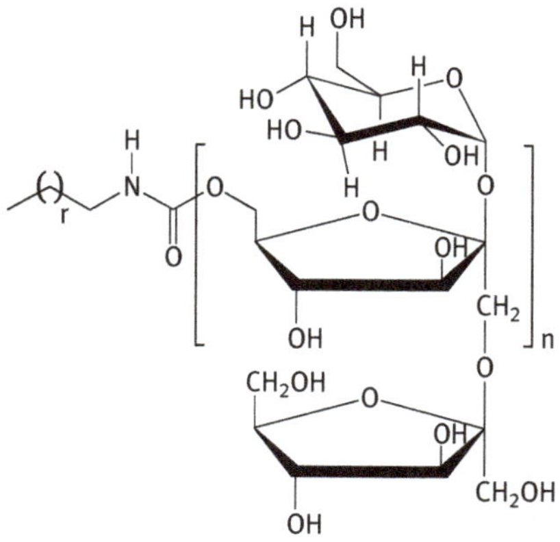

(GFn)

Fig. 2.5. Structure of INUTEC® SP1.

stabilizer. This copolymer is usually prepared by grafting a macromonomer such as methoxy polyethylene oxide methacrylate with polymethyl methacrylate. The "grafting onto" technique has also been used to synthesize polystyrene-polyethylene oxide graft copolymers. Recently, graft copolymers based on polysaccharides [25] have been developed for stabilization of disperse systems. One of the most useful graft copolymers is that based on inulin that is obtained from chicory roots. It is a linear polyfructose chain with a glucose end. When extracted from chicory roots, inulin has a wide range of chain lengths ranging from 2–65 fructose units. It is fractionated to obtain a molecule with narrow molecular weight distribution with a degree of polymerization > 23 and this is commercially available as INUTEC® N25. The latter molecule is used to prepare a series of graft copolymers by random grafting of alkyl chains (using alky isocyanate) onto the inulin backbone. The first molecule of this series is INUTEC® SP1 (Beneo-Remy, Belgium) that is obtained by random grafting of C_{12} alkyl chains. It has an average molecular weight of ~ 5000 Daltons and its structure is given in Fig. 2.5. The molecule is schematically illustrated below which shows the hydrophilic polyfructose chain (backbone) and the randomly attached alky chains. The main advantages of INUTEC® SP1 as stabilizer for disperse systems are:

1. Strong adsorption to the particle or droplet by multipoint attachment with several alkyl chains. This ensures lack of desorption and displacement of the molecule from the interface.
2. Strong hydration of the linear polyfructose chains both in water and in the presence of high electrolyte concentrations and high temperatures. This ensures effective steric stabilization.

References

[1]   Th. F. Tadros (ed.), *Surfactants*, Academic Press, London, 1984.
[2]   K. Holmberg, B. Jonsson, B. Kronberg and B. Lindman, *Surfactants and Polymers in Solution*, second edition, John Wiley and Sons, Ltd., Chichester, UK, 2003.
[3]   McCutcheon, *Detergents and Emulsifiers*, Allied Publishing Co., New Jersey, published annually.
[4]   N. M. van Os, J. R. Haak and L. A. M. Rupert, *Physico-chemical Properties of Selected Anionic, Cationic and Nonionic Surfactants*, Elsevier Publishing Co., Amsterdam, 1993.
[5]   M. R. Porter, *Handbook of Surfactants*, Chapman and Hall, Blackie, USA, 1994.
[6]   W. M. Linfield (ed.), *Anionic Surfactants*, Marcel Dekker, New York, 1967.
[7]   E. H. Lucasssen-Reynders, *Anionic Surfactants – Physical Chemistry of Surfactant Action*, Marcel Dekker, New York, 1981.
[8]   E. Jungermana, *Cationic Surfactants*, Marcel Dekker, New York, 1970.
[9]   N. Rubingh and P. M. Holland (eds.), *Cationic Surfactants – Physical Chemistry*, Marcel Dekker, New York, 1991.
[10] B. R. Buestein and C. L. Hiliton, *Amphoteric Surfactants*, Marcel Dekker, New York, 1982.
[11] M. J. Schick (ed.), *Nonionic Surfactants*, Marcel Dekker, New York, 1966.
[12] M. J. Schick (ed.), *Nonionic Surfactants: Physical Chemistry*, Marcel Dekker, New York, 1987.

[13] N. Schonfeldt, *Surface Active Ethylene Oxide Adducts*, Pergamon Press, USA, 1970.

[14] R. R. Zana and E. Alami, Gemini Surfactants, in: *Novel Surfactants*, K. Holberg (ed.), Chapter 12, Marcel Dekker, New York, 2003.

[15] W. von Rybinsky and K. Hill, in: *Novel Surfactants*, K. Holberg (ed.), Chapter 2, Marcel Dekker, New York, 2003.

[16] C. J. Drummond, C. Fong, I. Krodkiewska, B. J. Boyd and I. J. A. Baker, in: *Novel Surfactants*, K. Holberg (ed.), Chapter 3, Marcel Dekker, New York, 2003.

[17] B. W. Barry and G. M. T. Gray, *J. Colloid Interface Sci.*, **52**, 314 (1975).

[18] D. G. Oakenfull and L. R. Fisher, *J. Phys. Chem.*, **81**, 1838 (1977).

[19] D. M. Small, *Advan. Chem. Ser.*, **84**, 31 (1968).

[20] D. G. Oakenfull and L. R. Fisher, *J. Phys. Chem.*, **82**, 2443 (1978).

[21] R. Zana, *J. Phys. Chem.*, **82**, 2440 (1978).

[22] J. D. Desai and I. M. Banat, Microbial production of surfactants and their commercial potential, *Micobiol. Molec. Biol. Rev.* **61**, 47–64 (1997).

[23] K. K. Guatam and V. K. Tyagi, Microbial surfactants; A review, *J. Oleo. Sci.*, **55**, 155–166 (2006).

[24] A. J. J. Straathof and P. Aldercreutz (eds.) *Applied Biocatalysis*, Harwood Academic, Amsterdam, 2000.

[25] Th. Tadros, in: *Novel Surfactants*, K. Holberg (ed.), Chapter 16, Marcel Dekker, New York, 2003.

[26] B. Lindman, in: *Surfactants*, Th. F. Tadros (ed.), Academic Press, London, New York (1984). K. Holmberg, B. Jonsson, B. Kronberg and B. Lindman, *Surfactants and Polymers in Aqueous Solution*, 2nd Edition, John Wiley & Sons Ltd., USA, 2003.

3 Aggregation of surfactants, self-assembly structures, liquid crystalline phases

The physical properties of surface active agent solutions differ from those of non-amphipathic molecule solutions (such as sugars) in one major aspect, namely the abrupt changes in their properties above a critical concentration [1]. This is illustrated in Fig. 3.1 which shows plots of several physical properties (osmotic pressure, surface tension, turbidity, solubilization, magnetic resonance, equivalent conductivity and self-diffusion) as a function of concentration for an anionic surfactant. At low concentrations, most properties are similar to those of a simple electrolyte. One notable exception is the surface tension, which decreases rapidly with increasing surfactant concentration. However, all the properties (interfacial and bulk) show an abrupt change at a particular concentration, which is consistent with the fact that at and above this concentration surface active molecules or ions associate to form larger units. These associated units are called micelles (self-assembled structures) and the first-formed aggregates are generally approximately spherical in shape. A schematic representation of a spherical micelle is given in Fig. 3.2.

The concentration at which this association phenomenon occurs is known as the critical micelle concentration (cmc). Each surfactant molecules has a characteristic cmc value at a given temperature and electrolyte concentration. The most common technique for measuring the cmc is surface tension, γ, which shows a break at the cmc, after which γ remains virtually constant with further increases in concentration. How-

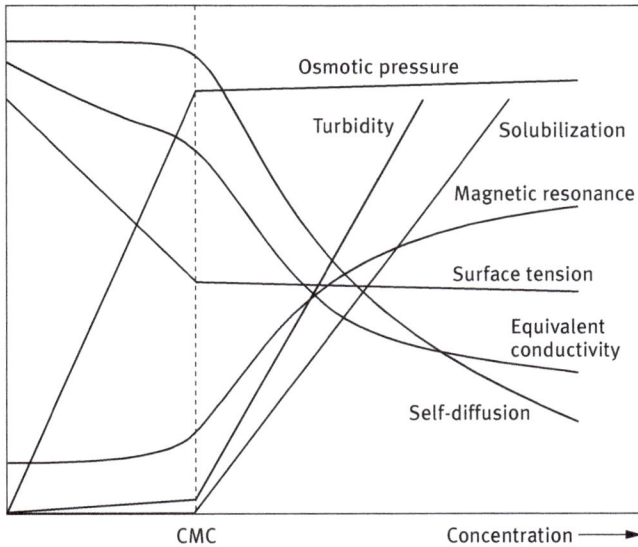

Fig. 3.1. Variation of solution properties with surfactant concentration.

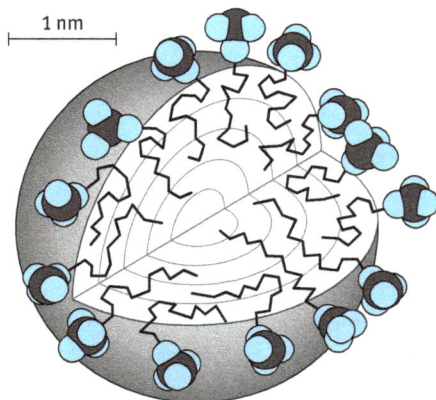

Fig. 3.2. Illustration of a spherical micelle for dodecyl sulfate [2].

ever, other techniques such as self-diffusion measurements, NMR and fluorescence spectroscopy can be applied. A compilation of cmc values was given in 1971 by Mukerjee and Mysels [3], which is clearly not an uptodate text, but is an extremely valuable reference. As an illustration, the cmc values of a number of surface active agents are given in Table 3.1, to show some of the general trends [3]. Within any class of surface active agent, the cmc decreases with increasing chain length of the hydrophobic portion (alkyl group). As a general rule, the cmc decreases by a factor of 2 for ionics (without added salt) and by a factor of 3 for nonionics on adding one methylene group to the alkyl chain. With nonionic surfactants, increasing the length of the hydrophilic group (polyethylene oxide) causes an increase in cmc.

Table 3.1. Surface active agent cmc/mol dm^{-3}

(A) Anionic	
Sodium octyl-l-sulfate	1.30×10^{-1}
Sodium decyl-l-sulfate	3.32×10^{-2}
Sodium dodecyl-l-sulfate	8.39×10^{-3}
Sodium tetradecyl-l-sulfate	2.05×10^{-3}
(B) Cationic	
Octyl trimethyl ammonium bromide	1.30×10^{-1}
Decetryl trimethyl ammonium bromide	6.46×10^{-2}
Dodecyl trimethyl ammonium bromide	1.56×10^{-2}
Hexactecyltrimethyl ammonium bromide	9.20×10^{-4}
(C) Nonionic	
Octyl hexaoxyethylene glycol monoether C_8E_6	9.80×10^{-3}
Decyl hexaoxyethylene glycol monoether $C_{10}E_6$	9.00×10^{-4}
Decyl nonaoxyethylene glycol monoether $C_{10}E_9$	1.30×10^{-3}
Dodecyl hexaoxyethylene glycol monoether $C_{12}E_6$	8.70×10^{-5}
Octylphenyl hexaoxyethylene glycol monoether C_8E_6	2.05×10^{-4}

In general, nonionic surfactants have lower cmc values than their corresponding ionic surfactants of the same alkyl chain length. Incorporation of a phenyl group into the alkyl group increases its hydrophobicity to a much smaller extent than increasing its chain length with the same number of carbon atoms. The valency of the counterion in ionic surfactants has a significant effect on the cmc. For example, increasing the valency of the counterion from 1 to 2 causes a reduction of the cmc by roughly a factor of 4.

The cmc is, to a first approximation, independent of temperature. This is illustrated in Fig. 3.3 which shows the variation of cmc of SDS with temperature. The cmc varies in a non-monotonic way by ca 10–20% over a wide temperature range. The shallow minimum around 25° C can be compared with a similar minimum in the solubility of hydrocarbon in water [4]. However, nonionic surfactants of the ethoxylate type show a monotonic decrease [4] of cmc with increasing temperature as illustrated in Fig. 3.3 for $C_{10}E_5$. The effect of addition of cosolutes, e.g. electrolytes and non-electrolytes, on the cmc can be very striking. For example, addition of 1:1 electrolyte to a solution of anionic surfactant gives a dramatic lowering of the cmc, which may amount to an order of magnitude. The effect is moderate for short-chain surfactants, but is much larger for long-chain ones. At high electrolyte concentrations, the reduction in cmc with an increase in the number of carbon atoms in the alkyl chain is much stronger than without added electrolyte. This rate of decrease at high electrolyte concentrations is comparable to that of nonionics. The effect of added electrolyte also depends on the valency of the added counterions. In contrast, for nonionics, addition of electrolytes causes only small variation in the cmc.

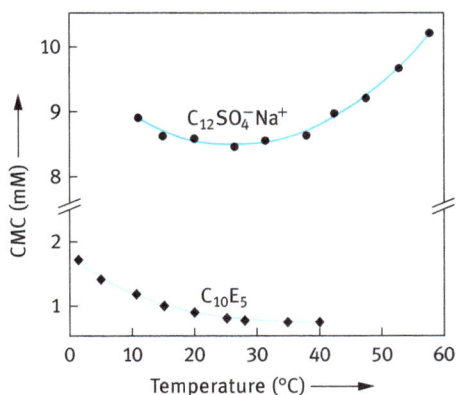

Fig. 3.3. Temperature dependence of the cmc of SDS and $C_{10}E_5$ [4].

Non-electrolytes such as alcohols can also cause a decrease in the cmc [5]. The alcohols are less polar than water and are distributed between the bulk solution and the micelles. The more preference they have for the micelles, the more they stabilize them.

A longer alkyl chain leads to a less favorable location in water and more favorable location in the micelles

The presence of micelles can account for many of the unusual properties of solutions of surface active agents. For example, it can account for the near constant surface tension value, above the cmc (see Fig. 3.1). It also accounts for the reduction in molar conductance of the surface active agent solution above the cmc, which is consistent with the reduction in mobility of the micelles as a result of counterion association. The presence of micelles also accounts for the rapid increase in light scattering or turbidity above the cmc. The presence of micelles was originally suggested by McBain [6] who suggested that below the cmc most of the surfactant molecules are unassociated, whereas in the isotropic solutions immediately above the cmc, micelles and surfactant ions (molecules) are thought to co-exist, the concentration of the latter changing very slightly as more surfactant is dissolved. However, the self-association of an amphiphile occurs in a stepwise manner with one monomer added to the aggregate at a time. For a long chain amphiphile, the association is strongly cooperative up to a certain micelle size where counteracting factors became increasingly important. Typically the micelles have a closely spherical shape in a rather wide concentration range above the cmc. Originally, it was suggested by Adam [7] and Hartley [8] that micelles are spherical in shape and have the following properties:

1. The association unit is spherical with a radius approximately equal to the length of the hydrocarbon chain.
2. The micelle contains about 50–100 monomeric units; aggregation number generally increases with increasing alkyl chain length.
3. With ionic surfactants, most counterions are bound to the micelle surface, thus significantly reducing the mobility from the value to be expected from a micelle with non-counterion bonding.
4. Micellization occurs over a narrow concentration range as a result of the high association number of surfactant micelles.
5. The interior of the surfactant micelle has essentially the properties of a liquid hydrocarbon. This is confirmed by the high mobility of the alkyl chains and the ability of the micelles to solubilize many water insoluble organic molecules, e.g. dyes and agrochemicals.

To a first approximation, micelles can, over a wide concentration range above the cmc, be viewed as microscopic liquid hydrocarbon droplets covered with polar head groups, which interact strongly with water molecules. It appears that the radius of the micelle core constituted of the alkyl chains is close to the extended length of the alkyl chain, i.e. in the range 1.5030 nm. As we will see later, the driving force for micelle formation is the elimination of contact between the alkyl chains and water. The larger a spherical micelle, the more efficient this is, since the volume-to-area ratio increases. It should be noted that the surfactant molecules in the micelles are not all extended. Only one molecule needs to be extended to satisfy the criterion that the radius of the micelle core is close to the extended length of the alkyl chain. The majority of surfactant molecules are in a disordered state. In other words, the interior of the micelle is close to that of the corresponding alkane in a neat liquid oil. This explains the large solubilization capacity of the micelle towards a broad range of non-polar and weakly polar substances. At the surface of the micelle associated counterions (in the region of 50–80% of the surfactant ions) are present. However, simple inorganic counterions are very loosely associated with the micelle. The counterions are very mobile (see below) and there is no specific complex formed with a definite counterion-head group distance. In other words, the counterions are associated by long-range electrostatic interactions.

A useful concept for characterizing micelle geometry is the critical packing parameter, CPP. The aggregation number N is the ratio between the micellar core volume, V_{mic}, and the volume of one chain, v,

$$N = \frac{V_{mic}}{v} = \frac{(4/3)\pi R_{mic}^3}{v} \tag{3.1}$$

where R_{mic} is the radius of the micelle.

The aggregation number, N, is also equal to the ratio of the area of a micelle, A_{mic}, to the cross-sectional area, a, of one surfactant molecule,

$$N = \frac{A_{mic}}{a} = \frac{4\pi R_{mic}^2}{a}. \tag{3.2}$$

Combining equations (3.1) and (3.2),

$$\frac{v}{R_{mic}a} = \frac{1}{3}. \tag{3.3}$$

Since R_{mic} cannot exceed the extended length of a surfactant alkyl chain, l_{max},

$$l_{max} = 1.5 + 1.265n_c. \tag{3.4}$$

This means that for a spherical micelle,

$$\frac{v}{l_{max}a} \leq \frac{1}{3}. \tag{3.5}$$

The ratio $v/(l_{max}a)$ is denoted as the critical packing parameter (CPP).

Although the spherical micelle model accounts for many of the physical properties of solutions of surfactants, a number of phenomena remain unexplained, without considering other shapes. For example, McBain [9] suggested the presence of two types of micelles, spherical and lamellar in order to account for the drop in molar conductance of surfactant solutions. The lamellar micelles are neutral and hence they account for the reduction in the conductance. Later, Harkins et al. [10] used McBain's model of lamellar micelles to interpret his X-ray results in soap solutions. Moreover, many modern techniques such as light scattering and neutron scattering

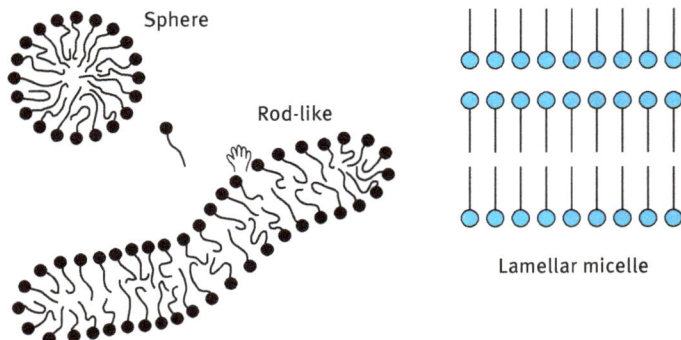

Fig. 3.4. Shapes of micelles.

indicate that in many systems the micelles are not spherical. For example, Debye and Anacker [11] proposed a cylindrical micelle to explain the light scattering results on hexadecyltrimethyl ammonium bromide in water. Evidence for disc-shaped micelles have also been obtained under certain conditions. A schematic representation of the spherical, lamellar and rod-shaped micelles, suggested by McBain, Hartley and Debye is given in Fig. 3.4. Many ionic surfactants show dramatic temperature-dependent solubility as illustrated in Fig. 3.5. The solubility first increases gradually with rising temperature, and then, above a certain temperature, there is a sudden increase in solubility with a further increase of temperature. The cmc increases gradually with increasing temperature. At a particular temperature, the solubility becomes equal to the cmc, i.e. the solubility curve intersects the cmc and this temperature is referred to as the Krafft temperature. At this temperature an equilibrium exists between slid-hydrated surfactant, micelles and monomers (i.e. the Krafft temperature is a "triple point"). Surfactants with ionic head groups and long straight alkyl chains have high Krafft temperatures. The Krafft temperature increases with the increase of the alkyl chain of the surfactant molecule. It can be reduced by introducing branching in the alkyl chain. The Krafft temperature is also reduced by using alkyl chains with wide

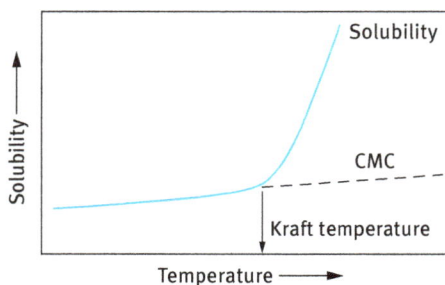

Fig. 3.5. Variation of solubility and critical micelle concentration (cmc) with temperature.

distribution of the chain length. Addition of electrolytes causes an increase in the Krafft temperature.

With nonionic surfactants of the ethoxylate type, an increase in temperature for a solution at a given concentration causes dehydration of the PEO chains and at a critical temperature the solution become cloudy. This is illustrated in Fig. 3.6 which shows the phase diagram of $C_{12}E_6$. Below the cloud point (CP) curve one can identify different liquid crystalline phases: Hexagonal – Cubic – Lamellar, which are schematically shown in Fig. 3.7.

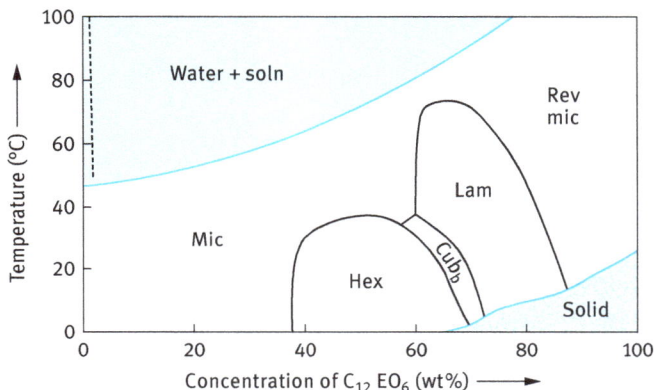

Fig. 3.6. Phase diagram of $C_{12}E_6$.

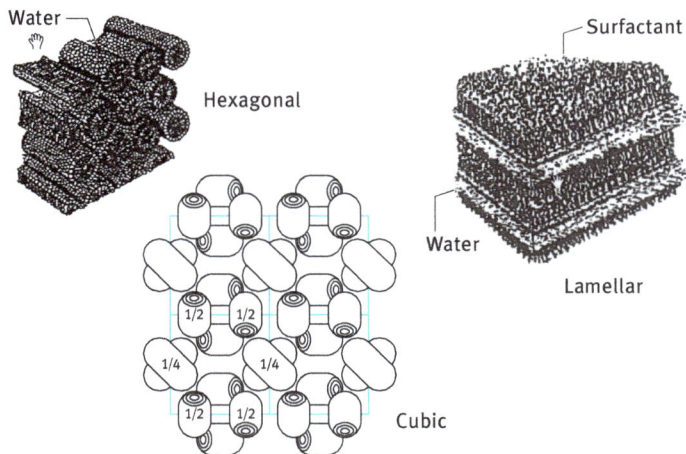

Fig. 3.7. Schematic picture of liquid crystalline phases.

3.1 Thermodynamics of micellization

The process of micellization is one of the most important characteristics of surfactant solution and hence it is essential to understand its mechanism (the driving force for micelle formation). This requires analysis of the dynamics of the process (i.e. the kinetic aspects) as well as the equilibrium aspects whereby the laws of thermodynamics may be applied to obtain the free energy, enthalpy and entropy of micellization.

3.1.1 Kinetic aspects

Micellization is a dynamic phenomenon in which n monomeric surfactant molecules associate to form a micelle S_n, i.e.,

$$nS \Leftrightarrow S_n \tag{3.6}$$

Hartley [8] envisaged a dynamic equilibrium whereby surface active agent molecules are constantly leaving the micelles whilst other molecules from solution enter the micelles. The same applies to the counterions with ionic surfactants, which can exchange between the micelle surface and bulk solution. Experimental investigation using fast kinetic methods such as stop flow, temperature and pressure jumps, and ultrasonic relaxation measurements have shown that there are two relaxation processes for micellar equilibrium [12–18] characterized by relaxation times τ_1 and τ_2. The first relaxation time, τ_1, is of the order of 10^{-7} s (10^{-8} to 10^{-3} s) and represents the lifetime of a surface active molecule in a micelle, i.e. it represents the association and dissociation rate for a single molecule entering and leaving the micelle, which may be represented by the equation,

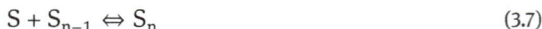

$$S + S_{n-1} \Leftrightarrow S_n \tag{3.7}$$

where K^+ and K^- represent the association and dissociation rate respectively for a single molecule entering or leaving the micelle.

The slower relaxation time τ_2 corresponds to a relatively slow process, namely the micellization-dissolution process represented by equation (3.6). The value of τ_2 is of the order of milliseconds ($10^{-3} - 1$ s) and hence can be conveniently measured by stopped flow methods. The fast relaxation time τ_1 can be measured using various techniques depending on its range. For example, τ_1 values in the range of $10^{-8}-10^{-7}$ s are accessible to ultrasonic absorption methods, whereas τ_1 in the range of $10^{-5}-10^{-3}$ s can be measured by pressure jump methods. The value of τ_1 depends on surfactant concentration, chain length and temperature. τ_1 increases with increasing chain length of surfactants, i.e. the residence time increases with increasing chain length.

The above discussion emphasizes the dynamic nature of micelles and it is important to realize that these molecules are in continuous motion and that there is a constant interchange between micelles and solution. The dynamic nature also applies to the counterions which exchange rapidly with lifetimes in the range $10^{-9}-10^{-8}$ s. Furthermore, the counterions appear to be laterally mobile and not to be associated with (single) specific groups on the micelle surfaces.

3.1.2 Equilibrium aspects: Thermodynamics of micellization

Various approaches have been employed in tackling the problem of micelle formation. The simplest approach treats micelles as a single phase, and this is referred to as the phase separation model. In this model, micelle formation is considered as a phase separation phenomenon and the cmc is then the saturation concentration of the amphiphile in the monomeric state whereas the micelles constitute the separated pseudophase. Above the cmc, a phase equilibrium exists with constant activity of the surfactant in the micellar phase. The Krafft point is viewed as the temperature at which solid hydrated surfactant, micelles and a solution saturated with undissociated surfactant molecules are in equilibrium at a given pressure.

Consider an anionic surfactant, in which n surfactant anions, S^-, and n counterions M^+ associate to form a micelle, i.e.,

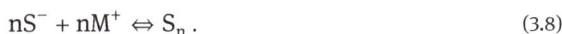

$$nS^- + nM^+ \Leftrightarrow S_n . \tag{3.8}$$

The micelle is simply a charged aggregate of surfactant ions plus an equivalent number of counterions in the surrounding atmosphere and is treated as a separate phase. The chemical potential of the surfactant in the micellar state is assumed to be constant at any given temperature, and this may be adopted as the standard chemical potential, μ_m^o, by analogy to a pure liquid or a pure solid. Considering the equilibrium between micelles and monomer, then,

$$\mu_m^o = \mu_1^o + RT \ln a, \tag{3.9}$$

where μ_1 is the standard chemical potential of the surfactant monomer and a_1 is its activity which is equal to $f_1 x_1$, where f_1 is the activity coefficient and x_1 the mole fraction. Therefore, the standard free energy of micellization per mol of monomer, ΔG_m^o, is given by

$$\Delta G_m^o = \mu_m^o - \mu_1^o = RT \ln a_1 \approx RT \ln x , \tag{3.10}$$

where f_1 is taken as unity (a reasonable value in very dilute solution). The cmc may be identified with x_1 so that

$$\Delta G_m^o = RT \ln[c.m.c]. \tag{3.11}$$

In equation (3.10), the cmc is expressed as a mole fraction, which is equal to $C/(55.5 + C)$, where C is the concentration of surfactant in mole dm^{-3}, i.e.,

$$\Delta G_m^o = RT \ln C - RT \ln(55.5 + C) . \tag{3.12}$$

It should be stated that ΔG° should be calculated using the cmc expressed as a mole fraction as indicated by equation (3.12). However, most cmc quoted in the literature are given in mole dm^{-3} and, in many cases ΔG° values have been quoted when the cmc was simply expressed in mol dm^{-3}. Strictly speaking, this is incorrect, since ΔG° should be based on x_1 rather than on C. The value of ΔG° when the cmc is expressed in mol dm^{-3} is substantially different from the ΔG° value when the cmc is expressed in a mole fraction. For example, for dodecyl hexaoxyethylene glycol the quoted cmc value is 8.7×10^{-5} mol dm^{-3} at 25° C. Therefore,

$$\Delta G^\circ = RT \ln \frac{8.7 \times 10^{-5}}{55.5 + 8.7 \times 10^{-5}} = -33.1 \text{ KJ mol}^{-1} \tag{3.13}$$

when the mole fraction scale is used. On the other hand,

$$\Delta G^\circ = RT \ln 8.7 \times 10^{-5} = -23.2 \text{ KJ mol}^{-1} \tag{3.14}$$

when the molarity scale is used.

The phase separation model has been questioned for two main reasons. Firstly, according to this model a clear discontinuity in the physical property of a surfactant solution, such as surface tension, turbidity, etc. should be observed at the cmc. This is not always found experimentally and the cmc is not a sharp break point. Secondly, if two phases actually exist at the cmc, then equating the chemical potential of the surfactant molecule in the two phases would imply that the activity of the surfactant in the aqueous phase would be constant above the cmc If this was the case, the surface tension of a surfactant solution should remain constant above the cmc. However, careful measurements have shown that the surface tension of a surfactant solution decreases slowly above the cmc, particularly when using purified surfactants.

A convenient solution for relating ΔG_m to [cmc] for ionic surfactants was given by Phillips [7] who arrived at the following expression,

$$\Delta G_m^\circ = \{2 - (p/n)\} RT \ln[\text{c.m.c.}], \tag{3.15}$$

where p is the number of free (unassociated) surfactant ions and n is the total number of surfactant molecules in the micelle. For many ionic surfactants, the degree of dissociation (p/n) is ~ 0.2 so that

$$\Delta G_m^\circ = 1.8 RT \ln[\text{c.m.c}]. \tag{3.16}$$

Comparison with equation (3.11) clearly shows that for similar ΔG_m, the [cmc] is about two orders of magnitude higher for ionic surfactants when compared with nonionic surfactant of the same alkyl chain length (see Table 3.1).

In the presence of excess added electrolyte, with mole fraction x, the free energy of micellization is given by the expression,

$$\Delta G_m^\circ = RT \ln[\text{c.m.c}] + \{1 - (p/n)\} \ln . \tag{3.17}$$

Equation (3.17) shows that as x increases, the [cmc] decreases.

It is clear from equation (3.15) that as p → 0, i.e. when most charges are associated with counterions,

$$\Delta G_m^o = 2RT \ln[c.m.c],$$ (3.18)

whereas when p ~ n, i.e. the counterions are bound to micelles,

$$\Delta G_m^o = RT \ln[c.m.c],$$ (3.19)

which is the same equation for nonionic surfactants.

3.2 Enthalpy and entropy of micellization

The enthalpy of micellization can be calculated from the variation of cmc with temperature. This follows from

$$-\Delta H^o = RT^2 \frac{d\ln[c.m.c]}{dT}.$$ (3.20)

The entropy of micellization can then be calculated from the relationship between ΔG^o and ΔH^o, i.e.,

$$\Delta G^o = \Delta H^o - T\Delta S^o.$$ (3.21)

Therefore ΔH^o may be calculated from the surface tension-log C plots at various temperatures. Unfortunately, the errors in locating the cmc (which in many cases is not a sharp point) leads to a large error in the value of ΔH^o. A more accurate and direct method of obtaining ΔH^o is microcalorimetry. As an illustration, the thermodynamic parameters, ΔG^o, ΔH^o, and $T\Delta S^o$ for octylhexaoxyethylene glycol monoether (C_8E_6) are given in Table 3.2.

Table 3.2. Thermodynamic quantities Δ for micellization of octylhexaoxyethylene glycol monoether.

Temp/° C	ΔG^o/KJ mol^{-1}	ΔH^o/KJ mol^{-1} (from cmc)	DeltaH^o/KJ mol l^{-1} (from calorimetry)	$T\Delta S^o$/KJ mol^{-1}
25	−21.3 ± 2.1	8.0 ± 4.2	20.1 ± 0.8	41.8 ± 1.0
40	−23.4 ± 2.1		14.6 ± 0.8	38.0 ± 1.0

It can be seen from Table 3.2 that ΔG^o is large and negative. However, ΔH^o is positive, indicating that the process is endothermic. In addition, $T\Delta S^o$ is large and positive which implies that in the micellization process there is a net increase in entropy. This positive enthalpy and entropy points to a different driving force for micellization from that encountered in many aggregation processes.

Table 3.3. Change of thermodynamic parameters of micellization of alkyl sulfoxide with increasing chain length of the alkyl group.

Surfactant	$\Delta G/KJ\ mol^{-1}$	$\Delta H^{\circ}/KJ\ mol^{-1}$	$T\Delta S^{\circ}/KJ\ mol^{-1}$
$C_6H_{13}S(CH_3)O$	−12.0	10.6	22.6
$C_7H_{15}S(CH_3)O$	−15.9	9.2	25.1
$C_8H_{17}S(CH_3)O$	−18.8	7.8	26.4
$C_9H_{19}S(CH_3)O$	−22.0	7.1	29.1
$C_{10}H_{21}S(CH_3)O$	−25.5	5.4	30.9
$C_{11}H_{23}S(CH_3)O$	−28.7	3.0	31.7

The influence of alkyl chain length of the surfactant on the free energy, enthalpy and entropy of micellization was demonstrated by Rosen [19] who listed these parameters as a function of alkyl chain length for sulfoxide surfactants. The results are given in Table 3.3 and it can be seen that the standard free energy of micellization becomes increasingly negative as the chain length increases. This is to be expected since the cmc decreases with increasing alkyl chain length. However, ΔH° becomes less positive and TΔS becomes more positive with increasing chain length of the surfactant. Thus, the large negative free energy of micellization is made up of a small positive enthalpy (which decreases slightly with increasing chain length of the surfactant) and a large positive entropy term $T\Delta S^{\circ}$, which becomes more positive as the chain is lengthened. As we will see in the next section, these results can be accounted for in terms of the hydrophobic effect which will be described in some detail.

3.3 Driving force for micelle formation

Until recently, the formation of micelles was regarded primarily as an interfacial energy process, analogous to the process of coalescence of oil droplets in an aqueous medium. If this was the case, micelle formation would be a highly exothermic process, as the interfacial free energy has a large enthalpy component. As mentioned above, experimental results have clearly shown that micelle formation involves only a small enthalpy change and is often endothermic. The negative free energy of micellization is the result of a large positive entropy. This led to the conclusion that micelle formation must be a predominantly entropy-driven process. Two main sources of entropy may have been suggested. The first is related to the so-called "hydrophobic effect". This effect was first established from a consideration of the free energy enthalpy and entropy of transfer of hydrocarbon from water to a liquid hydrocarbon. Some results are listed in Table 3.4. This table also lists the heat capacity change ΔC_p on transfer from water to a hydrocarbon, as well as $C_p^{o,gas}$, i.e. the heat capacity in the gas phase. It can be seen from Table 3.4 that the principal contribution to the value of ΔG° is the large positive

Table 3.4. Thermodynamic parameters for transfer of hydrocarbons from water to liquid hydrocarbon at 25 °C.

Hydrocarbon	ΔG^0 kJ mol^{-1}	ΔH^0 kJ mol^{-1}	ΔS^0 kJ mol^{-1}K^{-1}	ΔC_p^0 kJ mol^{-1}K^{-1}	$C_p^{0,gas}$ kJ mol^{-1}K^{-1}
C_2H_6	−16.4	10.5	88.2	−	−
C_3H_8	−20.4	7.1	92.4	−	−
C_4H_{10}	−24.8	3.4	96.6	−273	−143
C_5H_{12}	−28.8	2.1	105.0	−403	−172
C_6H_{14}	−32.5	0	109.2	−441	−197
C_6H_6	−19.3	−2.1	58.8	−227	−134
$C_6H_5CH_3$	−22.7	−1.7	71.4	−265	−155
$C_6H_5C_2H_5$	−26.0	−2.0	79.8	−319	−185
$C_6H_5C_3H_8$	−29.0	−2.3	88.2	−395	−

value of ΔS^0, which increases with increasing hydrocarbon chain length, whereas ΔH^0 is positive, or small and negative.

To account for this large positive entropy of transfer, several authors [20–22] suggest that the water molecules around a hydrocarbon chain are ordered, forming "clusters" or "icebergs". On transfer of an alkane from water to a liquid hydrocarbon, these clusters are broken, thus releasing water molecules which now have a higher entropy. This accounts for the large entropy of transfer of an alkane from water to a hydrocarbon medium. This effect is also reflected in the much higher heat capacity change on transfer, ΔC_p^0, when compared with the heat capacity in the gas phase, C_p^0. This effect is also operative on transfer of surfactant monomer to a micelle, during the micellization process. The surfactant monomers will also contain "structured" water around their hydrocarbon chain. On transfer of such monomers to a micelle, these water molecules are released and they have a higher entropy.

The second source of entropy increase on micellization may arise from the increase in flexibility of the hydrocarbon chains on their transfer from an aqueous to a hydrocarbon medium [20]. The orientations and bendings of an organic chain are likely to be more restricted in an aqueous phase compared to an organic phase. It should be mentioned that with ionic and zwitterionic surfactants, an additional entropy contribution, associated with the ionic head groups, must be considered. Upon partial neutralization of the ionic charge by the counterions when aggregation occurs, water molecules are released. This will be associated with an entropy increase which should be added to the entropy increase resulting from the hydrophobic effect mentioned above. However, the relative contribution of the two effects is difficult to assess in a quantitative manner.

3.4 Micellization in surfactant mixtures (mixed micelles)

In most industrial applications, more than one surfactant molecule is used in the formulation. It is, therefore, necessary to predict the types of possible interactions and whether these lead to some synergistic effects. Two general cases may be considered: surfactant molecules with no net interaction (with similar head groups) and systems with net interaction [1]. The first case is that when mixing two surfactants with the same head group but with different chain lengths. In analogy with the hydrophilic-lipophilic balance (HLB) for surfactant mixtures, one can also assume the cmc of a surfactant mixture (with no net interaction) to be an average of the two cmcs of the single components [1],

$$cmc = x_1\, cmc_1 + x_2\, cmc_2 \tag{3.22}$$

where x_1 and x_2 are the mole fractions of the respective surfactants in the system. However, the mole fractions should not be those in the whole system, but those inside the micelle. This means that equation (3.22) should be modified,

$$cmc = x_1^m\, cmc_1 + x_2^m\, cmc_2 \tag{3.23}$$

The superscript m indicates that the values are inside the micelle. If x_1 and x_2 are the solution composition, then,

$$\frac{1}{cmc} = \frac{x_1}{cmc_1} + \frac{x_2}{cmc_2}. \tag{3.24}$$

The molar composition of the mixed micelle is given by,

$$x_1^m = \frac{x_1\, cmc_2}{x_1\, cmc_2 + x_2\, cmc_1}. \tag{3.25}$$

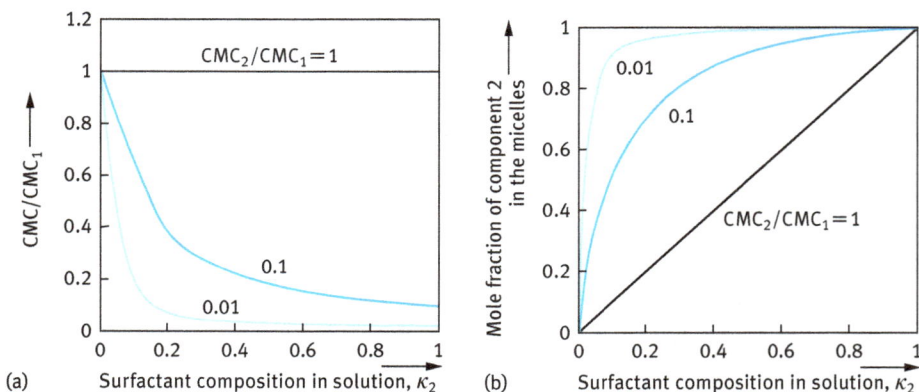

Fig. 3.8. Calculated cmc (a) and micellar composition (b) as a function of solution composition for three ratios of cmcs.

Fig. 3.8 shows the calculated cmc and the micelle composition as a function of solution composition using equations (3.24) and (.25) for three cases where $cmc_2/cmc_1 = 1$, 0.1 and 0.01. As can be seen, the cmc and micellar composition change dramatically with solution composition when the cmcs of the two surfactants vary considerably, i.e. when the ratio of cmcs is far from 1. This fact is used when preparing microemulsions where the addition of medium chain alcohol (like pentanol or hexanol) changes the properties considerably. If component 2 is much more surface active, i.e. $cmc_2/cmc_1 \ll 1$, and it is present in low concentrations (x_2 is of the order of 0.01), then from equation (3.25) $x_1^m \sim x_2^m \sim 0.5$, i.e. at the cmc of the systems the micelles are up to 50 % composed of component 2. This illustrates the role of contaminants in surface activity, e.g. dodecyl alcohol in sodium dodecyl sulfate (SDS).

Fig. 3.9 shows the cmc as a function of molar composition of the solution and in the micelles for a mixture of SDS and nonylphenol with 10 moles ethylene oxide ($NP-E_{10}$). If the molar composition of the micelles is used as the x-axis, the cmc is more or less the arithmetic mean of the cmcs of the two surfactants. If, on the other hand, the molar composition in the solution is used as the x-axis (which at the cmc is equal to the total molar concentration), then the cmc of the mixture shows a dramatic decrease at low fractions of $NP-E_{10}$. This decrease is due to the preferential absorption of $NP-E_{10}$ in the micelle. This higher absorption is due to the higher hydrophobicity of the $NP-E_{10}$ surfactant when compared with SDS.

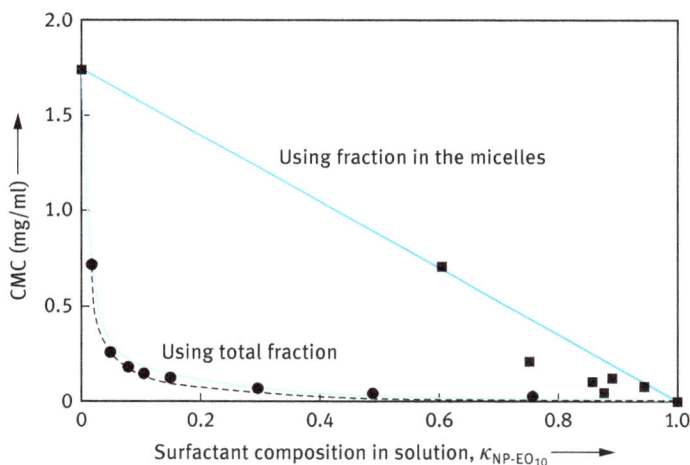

Fig. 3.9. Cmc as a function of surfactant composition, x_1, or micellar surfactant composition, x_1^m for the system SDS +NP-E_{10}.

With many industrial formulations, surfactants of different kinds are mixed together, for example anionics and nonionics. The nonionic surfactant molecules shield the repulsion between the negative head groups in the micelle and hence there will be a net interaction between the two types of molecules. Another example is the case when anionic and cationic surfactants are mixed, whereby very strong interaction will take place between the oppositely charged surfactant molecules. To account for this interaction, equation (3.25) has to be modified by introducing activity coefficients of the surfactants, f_1^m and f_2^m in the micelle,

$$cmc = x_1^m f_1^m \, cmc_1 + x_2^m \, f_2^m \, cmc_2 . \tag{3.26}$$

An expression for the activity coefficients can be obtained using the regular solutions theory [1],

$$\ln f_1^m = (x_2^m)^2 \beta \tag{3.27}$$

$$\ln f_2^m = (x_2^m)^2 \beta \tag{3.28}$$

where β is an interaction parameter between the surfactant molecules in the micelle. A positive β value means that there is a net repulsion between the surfactant molecules in the micelle, whereas a negative β value means a net attraction.

The cmc of the surfactant mixture and the composition x_1 are given by the following equations,

$$\frac{1}{cmc} = \frac{x_1}{f_1^m \, cmc_1} + \frac{x_2}{f_2^m \, cmc_2} \tag{3.29}$$

$$x_1^m = \frac{x_1 f_2^m \, cmc_2}{x_1 f_2^m \, cmc_2 + x_2 f_2^m \, cmc_1} . \tag{3.30}$$

Fig. 3.10 shows the effect of increasing the β parameter on the cmc and micellar composition for two surfactants with a cmc ratio of 0.1.

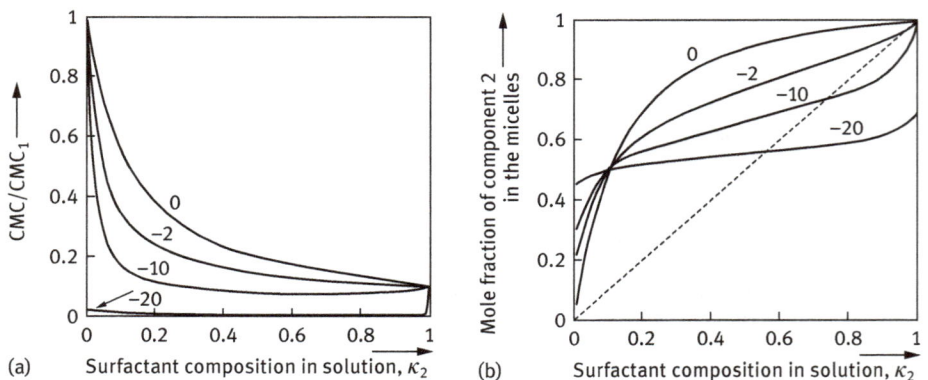

(a) Surfactant composition in solution, κ_2 (b) Surfactant composition in solution, κ_2

Fig. 3.10. Cmc (a) and micellar composition (b) for various values of β for a system with a cmc ratio cmc_2/cmc_1 of 0.1.

Fig. 3.10 shows that as β becomes more negative, the cmc of the mixture decreases. β values in the region of -2 are typical for anionic/nonionic mixtures, whereas values in the region of -10 to -20 are typical of anionic/cationic mixtures. With increasing the negative value of β, the mixed micelles tend towards a mixing ratio of $50:50$, which reflects the mutual electrostatic attraction between the surfactant molecules. The predicted cmc and micellar composition depends both on the ratio of the cmcs as well as the value of β. When the cmcs of the single surfactants are similar, the predicted value of the cmc is very sensitive to small variations in β. On the other hand, when the ratio of the cmcs is large, the predicted value of the mixed cmc and the micellar composition are insensitive to variation of the β parameter. For mixtures of nonionic and ionic surfactants, the β decreases with increasing electrolyte concentration. This is due to the screening of the electrostatic repulsion on the addition of electrolyte. With some surfactant mixtures, the β decreases with increasing temperature, i.e. the net attraction decreases with increasing temperature.

3.5 Surfactant self-assembly

Surfactant micelles and bilayers are the building blocks of most self-assembly structures. One can divide the phase structures into two main groups [1]:
1. those that are built of limited or discrete self-assemblies, which may be characterized roughly as spherical, prolate or cylindrical;
2. infinite or unlimited self-assemblies whereby the aggregates are connected over macroscopic distances in one, two or three dimensions.

The hexagonal phase (see below) is an example of one-dimensional continuity, the lamellar phase of two-dimensional continuity, whereas the bicontinuous cubic phase and the sponge phase (see later) are examples of three-dimensional continuity. These two types are schematically illustrated in Fig. 3.11.

3.5.1 Structure of liquid crystalline phases

The above-mentioned unlimited self-assembly structures in 1D, 2D or 3D are referred to as liquid crystalline structures. The latter behave as fluids and are usually highly viscous. At the same time, X-ray studies of these phases yield a small number of relatively sharp lines which resemble those produced by crystals [1]. Since they are fluids they are less ordered than crystals, but because of the X-ray lines and their high viscosity it is also apparent that they are more ordered than ordinary liquids. Thus, the term liquid crystalline phase is very appropriate for describing these self-assembled structures. Below a brief description of the various liquid crystalline structures that

Discrete

Connected

ID   2D   3D

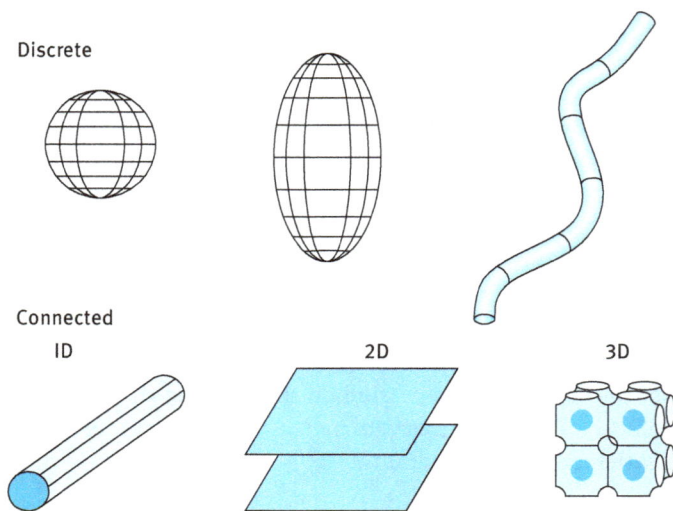

Fig. 3.11. Schematic representation of self-assembly structures.

can be produced with surfactants is given and Table 3.5 shows the most commonly used notation to describe these systems.

Table 3.5. Notation of the most common liquid crystalline structures.

Phase Structure	Abbreviation	Notation
Micellar	mic	L_1, S
Reversed micellar	rev mic	L_2, S
Hexagonal	hex	H_1, E, M_1, middle
Reversed hexagonal	rev hex	H_2, F, M_2
Cubic (normal micellar)	cub_m	I_1, S_{1c}
Cubic (reversed micelle)	cub_m	I_2
Cubic (normal bicontinuous)	cub_b	I_1, V_1
Cubic (reversed bicontinuous)	cub_b	I_2, V_2
Lamellar	lam	L_α, D, G, neat
Gel	gel	L_β
Sponge phase (reversed)	spo	L_3 (normal), L_4

3.5.2 Hexagonal phase

This phase is built up of (infinitely) long cylindrical micelles arranged in a hexagonal pattern, with each micelle being surrounded by six other micelles, as schematically shown in Fig. 3.7. The radius of the circular cross section (which may be somewhat deformed) is again close to the surfactant molecule length [1].

3.5.3 Micellar cubic phase

This phase is built up of regular packing of small micelles, which have similar properties of small micelles in the solution phase. However, the micelles are short prolates (axial ratio 1–2) rather than spheres since this allows a better packing. The micellar cubic phase is highly viscous. A schematic representation of the micellar cubic phase [1] is shown in Fig. 3.7.

3.5.4 Lamellar phase

This phase is built of bilayers of surfactant molecules alternating with water layers. The thickness of the bilayers is somewhat lower than twice the surfactant molecule length. The thickness of the water layer can vary over wide ranges, depending on the nature of the surfactant. The surfactant bilayer can range from being stiff and planar to being very flexible and undulating. A schematic representation of the lamellar phase [1] is shown in Fig. 3.7.

3.5.5 Bicontinuous cubic phases

These phases can be a number of different structures where the surfactant molecules form aggregates that penetrate space, forming a porous connected structure in three dimensions. They can be considered as structures formed by connecting rod-like micelles (branched micelles), or bilayer structures [1] as illustrated in Fig. 3.12.

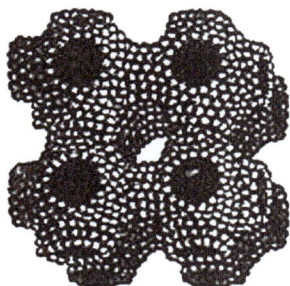

Fig. 3.12. Bicontinuous structure with the surfactant molecules aggregated into connected films characterized by two curvatures of opposite sign [9].

3.5.6 Reversed structures

Except for the lamellar phase, which is symmetrical around the middle of the bilayer, the different structures have a reversed counterpart in which the polar and non-polar parts have changed roles. For example, a hexagonal phase is built up of hexagonally packed water cylinders surrounded by the polar head groups of the surfactant molecules and a continuum of the hydrophobic parts. Similarly, reversed (micellar-type) cubic phases and reversed micelles consist of globular water cores surrounded by surfactant molecules. The radii of the water cores are typically in the range 2–10 nm.

3.6 Experimental studies of the phase behavior of surfactants

One of the earliest (and qualitative) techniques for identification of the different phases is the use of polarizing microscopy. This is based on the scattering of normal and polarized light which differs for isotropic (such as the cubic phase) and anisotropic (such as the hexagonal and lamellar phases) structures. Isotropic phases are clear and transparent, while anisotropic liquid crystalline phases scatter light and appear more or less cloudy. Using polarized light and viewing the samples through cross polarizers gives a black picture for isotropic phases, whereas anisotropic ones give bright images. The patterns in a polarization microscope are distinctly different for different anisotropic phases and can therefore be used to identify the phases, e.g. to distinguish between hexagonal and lamellar phases [23]. A typical optical micrograph for the hexagonal and lamellar phases (obtained using polarizing microscopy) is shown in Fig. 3.13. The hexagonal phase shows a "fan-like" appearance, whereas the lamellar phase shows "oily streaks" and "Maltese crosses".

(a)   (b)

Fig. 3.13. Texture of the hexagonal (a) and lamellar phase (b) obtained using polarizing microscopy.

Another qualitative method is to measure the viscosity as a function of surfactant concentration. The cubic phase is very viscous and often quite stiff and it appears as a clear "gel". The hexagonal phase is less viscous than the cubic phase and the lamellar phase is much less viscous than the cubic phase. However, viscosity measurement does not allow an unambiguous determination of the phases in the sample.

Most qualitative techniques for identification of the various liquid crystalline phases are based on diffraction studies, either light, X-ray or neutron. The liquid crystalline structures have a repetitive arrangement of aggregates and observation of a diffraction pattern can give evidence of long-range order and it can distinguish between alternative structures.

Another very useful technique to identify the different phases is NMR spectroscopy. One observes the quadropole splittings in deuterium NMR [24]. This is illustrated in Fig. 3.14. For isotropic phases such as micelles, cubic and sponge phases one observes a narrow singlet (Fig. 3.14a). For a single isotropic phase such as hexagonal or lamellar structures, a doublet is obtained (Fig. 3.14b).

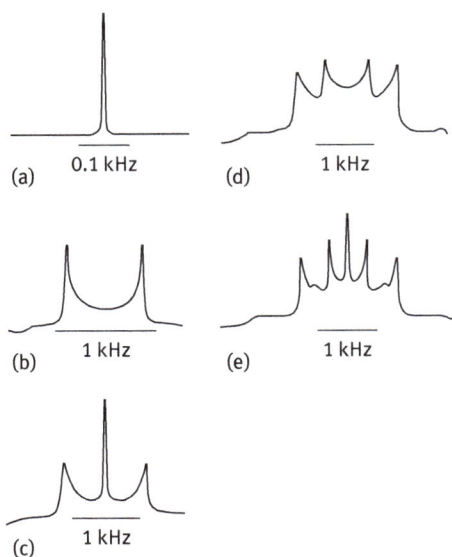

(a) 0.1 kHz
(b) 1 kHz
(c) 1 kHz
(d) 1 kHz
(e) 1 kHz

Fig. 3.14. ^{2}H NMR spectra of surfactants in heavy water (D_2O).

The magnitude of the "splitting" depends on the type of liquid crystalline phase, which is twice as much for the lamellar phase when compared with the hexagonal phase. For one isotropic and one anisotropic phase, one obtains one singlet and one doublet (Fig. 3.14c). For two anisotropic phases (lamellar and hexagonal) one observes two doublets (Fig. 3.14d). In a three-phase region with two anisotropic phases and one isotropic phase, one observes two doublets and one singlet (Fig. 3.14e).

The distinction between normal and reversed phases can be easily carried out using conductivity measurements. For normal phases which are "water rich", the conductivity is high. In contrast, for reversed phases that are "water poor", the conductivity is much lower (by several orders of magnitude).

References

[1] B. Lindman, in: *Surfactants*, Tadros, Th.F. (ed.),Academic Press, London, New York, 1984.
     K. Holmberg, B. Jonsson, B. Kronberg and B. Lindman, *Surfactants and Polymers in Aqueous Solution*, 2nd edition, John Wiley & Sons Ltd., USA, 2003.
[2] J. Istraelachvili, *Intermolecular and Surface Forces, with Special applications to Colloidal and Biological Systems*, Academic Press, London, 1985, p 251.
[3] P. Mukerjee and K. J. Mysels, *Critical Micelle Concentrations of Aqueous Surfactant Systems*, National Bureau of Standards Publication, Washington D.C., USA, 1971.
[4] P. H. Elworthy, A. T. Florence and C. B. Macfarlane, *Solubilization by Surface Active Agents*, Chapman and Hall, London, 1968.
[5] K. Shinoda, T. Nagakawa, B. I. Tamamushi and T. Isemura, *Colloidal Surfactants, Some Physicochemical Properties*, Academic Press, London, 1963.
[6] J. W. McBain, *Trans. Faraday Soc.*, **9**, 99 (1913).
[7] N. K. Adam, *J. Phys. Chem.*, **29**, 87 (1925).
[8] G. S. Hartley, *Aqueous Solutions of Paraffin Chain Salts*, Hermann and Cie, Paris, 1936.
[9] J. W. McBain, *Colloid Science*, Heath, Boston, 1950.
[10] W. D. Harkins, W. D. Mattoon and M. L. Corrin, *J. Amer. Chem. Soc.*, **68**, 220 (1946); *J. Colloid Sci.*, **1**, 105 (1946)
[11] P. Debye, and E. W. Anaker, *J. Phys. and Colloid Chem.*, **55**, 644 (1951).
[12] E. A. G. Anainsson and S. N. Wall, *J. Phys. Chem.*, **78**, 1024 (1974); **79**, 857 (1975).
[13] E. A. G. Aniansson, S. N. Wall, M. Almagren, H. Hoffmann, W. Ulbricht, R. Zana, J. Lang and C. Tondre, *J. Phys. Chem.*, **80**, 905 (1976).
[14] J. Rassing, P. J. Sams and E. Wyn-Jones, *J. Chem. Soc., Faraday II*, **70**, 1247 (1974).
[15] M. J. Jaycock and R. H. Ottewill, *Fourth Int. Congress Surface Activity*, **2**, 545 (1964).
[16] T. Okub, H. Kitano, T. Ishiwatari, and N. Isem, *Proc. Royal Soc.*, **A36**, 81 (1979).
[17] J. N. Phillips, *Trans. Faraday Soc.*, **51**, 561 (1955).
[18] M. Kahlweit and M. Teubner, *Adv. Colloid Interface Sci.*, **13**, 1 (1980).
[19] M. L. Rosen, *Surfactants and Interfacial Phenomena*, Wiley-Interscience, New York, 1978.
[20] C. Tanford, *The Hydrophobic Effect*, 2nd edn., Wiley, New York, 1980.
[21] G. Stainsby and A. E. Alexander, *Trans. Faraday Soc.*, **46**, 587 (1950).
[22] R. H. Arnow and L. Witten, *J. Phys. Chem.*, **64**, 1643 (1960).
[23] F. B. Rosevaar, *J. Soc. Cosmet. Chem.*, **19**, 581 (1968).
[24] A. Khan, K. Fontell, G. Lindblom and B. Lindman, *J. Phys. Chem.*, **86**, 4266 (1982).

4 Surfactant adsorption at interfaces

4.1 Introduction

Surfactants play a major role in the formulation of most chemical products. They are used for stabilization of emulsions, nanoemulsions, microemulsions and suspensions. Secondly, surfactants are added in emulsifiable concentrates for their spontaneous dispersion on dilution. The surfactant needs to accumulate at the interface, a process that is generally described as adsorption. The simplest interface is that of air/liquid and in this case the surfactant will adsorb with the hydrophilic group pointing towards the polar liquid (water) leaving the hydrocarbon chain pointing towards the air. This process results in lowering of the surface tension γ. Typically, surfactants show a gradual reduction in γ, until the critical micelle concentration (cmc) is reached above which the surface tension remains virtually constant. Hydrocarbon surfactants of the ionic, nonionic or zwitterionic ionic type lower the surface tension to limiting values reaching $30-40$ mNm^{-1} depending on the nature of the surfactant. Lower values may be achieved using fluorocarbon surfactants, typically of the order of 20 mNm^{-1}. It is, therefore, essential to understand the adsorption and conformation of surfactants at the air/liquid interface.

With emulsions, nanoemulsions and microemulsion, the surfactant adsorbs at the oil/water interface, with the hydrophilic head group immersed in the aqueous phase, leaving the hydrocarbon chain in the oil phase. Again, the mechanism of stabilization of emulsions, nano-emulsions and microemulsions depends on the adsorption and orientation of the surfactant molecules at the liquid/liquid interface. Surfactants consist of a small number of units and they are mostly reversibly adsorbed, allowing one to apply some thermodynamic treatments. In this case it is possible to describe the adsorption in terms of various interaction parameters such as chain/surface, chain solvent and surface solvent. Moreover, the configuration of the surfactant molecule can be simply described in terms of these possible interactions.

The use of surfactants (ionic, nonionic and zwitterionic) to control the stability behavior of suspensions is of considerable technological importance. Surfactants are used in the formulation of dyestuffs, paints, paper coatings, agrochemicals, pharmaceuticals, ceramics, printing inks, etc. They are a particularly robust form of stabilization, which is useful at high disperse volume fractions and high electrolyte concentrations, as well as under extreme conditions of high temperature, pressure and flow. In particular, surfactants are essential for the stabilization of suspensions in nonaqueous media, where electrostatic stabilization is less successful. The key to understanding how surfactants function as stabilizers is to know their adsorption and conformation at the solid/liquid interface.

4.2 Adsorption of surfactants at the air/liquid (A/L) and liquid/liquid (L/L) interfaces

Before describing surfactant adsorption at the air/liquid (A/L) and liquid/liquid (L/L) interface it is essential to define the interface. The surface of a liquid is the boundary between two bulk phases, namely liquid and air (or the liquid vapor). Similarly, an interface between two immiscible liquids (oil and water) may be defined providing a dividing line is introduced since the interfacial region is not a layer that is one molecule thick, but usually has a thickness δ with properties that are different from the two bulk phases α and β [1]. However, Gibbs [2] introduced the concept of a mathematical dividing plane Z_σ in the interfacial region (Fig. 4.1).

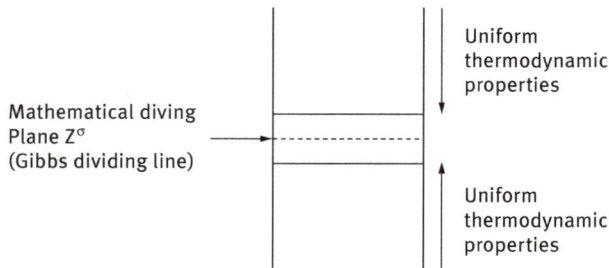

Fig. 4.1. Gibbs convention for an interface.

In this model the two bulk phases α and β are assumed to have uniform thermodynamic properties up to Z_σ. This picture applies for both the air/liquid and liquid/liquid interface (with A/L interfaces, one of the phase is air saturated with the vapor of the liquid).

   Using the Gibbs model, it is possible to obtain a definition of the surface or interfacial tension γ, starting from the Gibbs–Deuhem equation [2], i.e.,

$$dG^\sigma = -s^\sigma\,dt + A\,d\gamma + \Sigma n_1\,d\mu_i, \qquad (4.1)$$

where G^σ is the surface free energy, S^σ is the entropy, A is the area of the interface, n_i is the number of moles of component i with chemical potential μ_i at the interface. At constant temperature and composition of the interface (i.e. in absence of any adsorption),

$$\gamma = \left(\frac{\partial G^\sigma}{\partial A}\right)_{T,n_i}. \qquad (4.2)$$

It is obvious from equation (4.2) that for a stable interface γ should be positive. In other words, the free energy should increase if the area of the interface increases, otherwise the interface will become convoluted, increasing the interfacial area, until the liquid evaporates (for A/L case) or the two "immiscible" phases dissolved in each other (for the L/L case).

It is clear from equation (4.2) that surface or interfacial tension, i.e. the force per unit length tangentially to the surface measured in units of mNm^{-1}, is dimensionally equivalent to an energy per unit area measured in mJm^{-2}. For this reason, it has been stated that the excess surface free energy is identical to the surface tension, but this is true only for a single component system, i.e. a pure liquid (where the total adsorption is zero).

There are generally two approaches for treating surfactant adsorption at the A/L and L/L interface. The first approach, adopted by Gibbs, treats adsorption as an equilibrium phenomenon whereby the second law of thermodynamics may be applied using surface quantities. The second approach, referred to as the equation of state approach, treats the surfactant film as a two-dimensional layer with a surface pressure π that may be related to the surface excess Γ (amount of surfactant adsorbed per unit area). Below, these two approaches are summarized.

4.2.1 The Gibbs adsorption isotherm

Gibbs [2] derived a thermodynamic relationship between the surface or interfacial tension γ and the surface excess Γ (adsorption per unit area). The starting point of this equation is the Gibbs–Deuhem equation given above (equation (4.1)). At equilibrium (where the rate of adsorption is equal to the rate of desorption), $dG^\sigma = 0$. At constant temperature, but in the presence of adsorption,

$$dG^\sigma = -S^\sigma dT + A\,d\gamma + \sum n_i\,d\mu_i$$

or

$$d\gamma = -\sum \frac{n_i^\sigma}{A}\,d\mu_i = -\sum \Gamma_i\,d\mu_i \tag{4.3}$$

where $\Gamma_i = n_i^\sigma/A$ is the number of moles of component i and adsorbed per unit area.

Equation (4.3) is the general form for the Gibbs adsorption isotherm. The simplest case of this isotherm is a system of two components in which the solute (2) is the surface active component, i.e. it is adsorbed at the surface of the solvent (1). For such a case, equation (4.3) may be written as,

$$-d\gamma = \Gamma_1^\sigma d\mu_1 + \Gamma_2^\sigma d\mu_2 \tag{4.4}$$

and if the Gibbs dividing surface is used, $\Gamma_1 = 0$ and,

$$-d\gamma = \Gamma_{1,2}^\sigma d\mu_2 \tag{4.5}$$

where $\Gamma_{2,1}^\sigma$ is the relative adsorption of (2) with respect to (1). Since,

$$\mu_2 = \mu_2^o + RT \ln a_2^L \tag{4.6}$$

or,

$$d\mu_2 = RT\,d\ln a_2^L, \tag{4.7}$$

then,

$$-dy = \Gamma_{2,1}^{\sigma}\,RT\,d\ln a_2^L \tag{4.8}$$

or

$$\Gamma_{2,1}^{\sigma} = -\frac{1}{RT}\left(\frac{dy}{d\ln a_2^L}\right) \tag{4.9}$$

where a_2^L is the activity of the surfactant in bulk solution that is equal to $C_2 f_2$ or $x_2 f_2$, where C_2 is the concentration of the surfactant in moles dm^{-3} and x_2 is its mole fraction.

Equation (4.9) allows one to obtain the surface excess (abbreviated as Γ_2) from the variation of surface or interfacial tension with surfactant concentration. Note that $a_2 \sim C_2$, since in dilute solutions $f_2 \sim 1$. This approximation is valid since most surfactants have low cmc (usually less than 10^{-3} mol dm^{-3}) and adsorption is complete at or just below the cmc.

The surface excess Γ_2 can be calculated from the linear portion of the $y - \log C_2$ curves before the cmc. Such $y - \log C$ curves are illustrated in Fig. 4.2 for the air/water and o/w interfaces; $[C_{SAA}]$ denotes the concentration of surface active agent in bulk solution. It can be seen that for the A/W interface y decreases from the value for water (72 mNm^{-1} at 20°C) reaching about 25–30 mNm^{-1} near the cmc. This is clearly schematic since the actual values depend on the surfactant nature. For the o/w case, y decreases from a value of about 50 mNm^{-1} (for a pure hydrocarbon-water interface) to ~1–5 mNm^{-1} near the cmc (again depending on the nature of the surfactant).

As mentioned above, Γ_2 can be calculated from the slope of the linear position of the curves shown in Fig. 4.2 just before the cmc is reached. From Γ_2, the area per surfactant ion or molecule can be calculated since

$$\text{Area/molecule} = \frac{1}{\Gamma_2 N_{av}}, \tag{4.10}$$

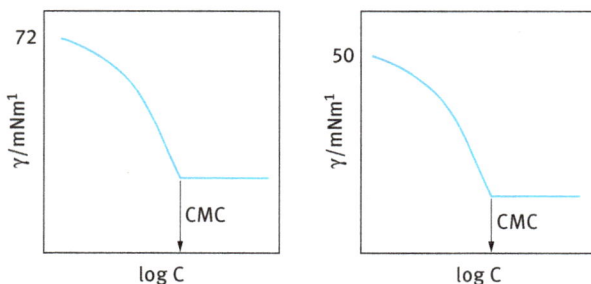

Fig. 4.2. Variation of surface and interfacial tension with log $[C_{SAA}]$ at the air/water and oil/water interface.

where N_{av} is the Avogadro's constant. Determining the area per surfactant molecule is very useful since it gives information on surfactant orientation at the interface. For example, for ionic surfactants such as alkyl sulfates, the area per surfactant is determined by the area occupied by the alkyl chain and head group if these molecules lie flat at the interface. In this case the area per molecule increases with increasing alkyl chain length. For vertical orientation, the area per surfactant ion is determined by that occupied by the charged head group, which at low electrolyte concentration will be in the region of $0.40 \, nm^2$. Such an area is larger than the geometrical area occupied by a sulfate group, as a result of the lateral repulsion between the head group. On addition of electrolytes, this lateral repulsion is reduced and the area/surfactant ion for vertical orientation will be lower than $0.4 \, nm^2$ (reaching in some cases $0.2 \, nm^2$).

Another important point can be made from the $\gamma - \log C$ curves. At concentration just before the break point, one has the condition of constant slope, which indicates that saturation adsorption has been reached.

$$\left(\frac{\partial \gamma}{\partial \ln a_2} \right)_{p,T} = \text{constant.} \tag{4.11}$$

Just above the break point,

$$\left(\frac{\partial \gamma}{\partial \ln a_2} \right)_{p,T} = 0, \tag{4.12}$$

indicating the constancy of γ with $\log C$ above the cmc. Integration of equation (4.12) gives

$$\gamma = \text{constant} \times \ln a_2 . \tag{4.13}$$

Since γ is constant in this region, then a_2 must remain constant. This means that addition of surfactant molecules, above the cmc must result in association to form units (micellar) with low activity.

As mentioned before, the hydrophilic head group may be un-ionized, e.g. alcohols or poly(ethylene oxide) alkane or alkyl phenol compounds, weakly ionized such as carboxylic acids or strongly ionized such as sulfates, sulfonates and quaternary ammonium salts. The adsorption of these different surfactants at the air/water and oil/water interface depends on the nature of the head group. With nonionic surfactants, repulsion between the head groups is small and these surfactants are usually strongly adsorbed at the surface of water from very dilute solutions. Nonionic surfactants have much lower cmc values when compared with ionic surfactants with the same alkyl chain length. Typically, the cmc is in the region of 10^{-5}–$10^{-4} \, mol \, dm^{-3}$. Such nonionic surfactants form closely packed adsorbed layers at concentrations lower than their cmc values. The activity coefficient of such surfactants is close to unity and is only slightly affected by addition of moderate amounts of electrolytes (or change in the pH of the solution). Thus, nonionic surfactant adsorption is the simplest case since the solutions can be represented by a two component system and the adsorption can be accurately calculated using equation (4.9).

With ionic surfactants, on the other hand, the adsorption process is relatively more complicated since one has to consider the repulsion between the head groups and the effect of presence of any indifferent electrolyte. Moreover, the Gibbs adsorption equation has to be solved taking into account the surfactant ions, the counterion and any indifferent electrolyte ions present. For a strong surfactant electrolyte such as an $Na^+ R^-$,

$$\Gamma_2 = \frac{1}{2RT} \frac{\partial \gamma}{\partial \ln a+}. \tag{4.14}$$

The factor of 2 in equation (4.14) arises because both surfactant ion and counterion must be adsorbed to maintain neutrally, and $d\gamma/d\ln a\pm$ is twice as large as for an unionized surfactant.

If a non-adsorbed electrolyte, such as NaCl, is present in large excess, then any increase in concentration of Na^+R^- produces a negligible increase in Na^+ ion concentration and therefore $d\mu_{Na}$ becomes negligible. Moreover, $d\mu_{Cl}$ is also negligible, so that the Gibbs adsorption equation reduces to

$$\Gamma_2 = -\frac{1}{RT} \left(\frac{\partial \gamma}{\partial \ln C_{NaR}} \right), \tag{4.15}$$

i.e. it becomes identical to that for a nonionic surfactant.

The above discussion clearly illustrates that for calculation of Γ_2 from the $\gamma - \log C$ curve one has to consider the nature of the surfactant and the composition of the medium. For nonionic surfactants the Gibbs adsorption equation (4.9) can be directly used. For ionic surfactant, in absence of electrolytes the right-hand side of the equation (4.9) should be divided by 2 to account for surfactant dissociation. This factor disappears in the presence of a high concentration of an indifferent electrolyte.

The surface excess concentration at surfactant saturation Γ_m is a useful measure of the effectiveness of adsorption at the A/L and L/L interface, since it is the maximum value that adsorption can attain. The effectiveness of adsorption is an important factor in determining such properties as foaming, emulsification and wetting. Tightly packed, coherent surfactant films have very different interfacial properties than loosely packed, noncoherent films. As mentioned before, for surfactants with singly hydrophilic group, either ionic or nonionic, the area occupied by a surfactant molecule at the interface appears to be determined by the area occupied by the hydrated hydrophilic group rather than by the hydrophobic group. If a second hydratable, hydrophilic group is introduced in the molecule, that portion of the molecule between the two hydrophilic groups tends to lie flat at the interface, and the area occupied by the molecule at the interface is increased.

4.2.2 Equation of state approach

In this approach, one relates the surface pressure π with the surface excess Γ_2. The surface pressure is defined by the equation,

$$\pi = \gamma_0 - \gamma \qquad (4.16)$$

where γ_0 is the surface or interfacial tension before adsorption and γ that after adsorption.

For an ideal surface film, behaving as a two-dimensional gas, the surface pressure π is related to the surface excess Γ_2 by the equation,

$$\pi A = n_2 RT \qquad (4.17)$$

or

$$\pi = (n_2/A)RT = \Gamma_2 RT . \qquad (4.18)$$

Differentiating equation (4.17) at constant temperature,

$$d\pi = RT\, d\Gamma_2 . \qquad (4.19)$$

Using the Gibbs equation,

$$d\pi = -d\gamma = \Gamma_2 RT\, d\ln a_2 \approx \Gamma_2 RT\, d\ln C_2 . \qquad (4.20)$$

Combining equations (4.19) and (4.20)

$$d\ln \Gamma_2 = d\ln C_2 \qquad (4.21)$$

or

$$\Gamma_2 = K C_2^\alpha. \qquad (4.22)$$

Equation (4.22) is referred to as the Henry's law isotherm which predicts a linear relationship between Γ_2 and C_2.

It is clear that equations (4.16) and (4.19) are based on an idealized model in which the lateral interaction between the molecules has not been considered. Moreover, in this model the molecules are considered to be dimensionless. This model can only be applied at very low surface coverage where the surfactant molecules are so far apart that lateral interaction may be neglected. Moreover, under these conditions the total area occupied by the surfactant molecules is relatively small compared to the total interfacial area.

At significant surface coverage, the above equations have to be modified to take into account both lateral interaction between the molecules as well as the area occupied by them. Lateral interaction may reduce π if there is attraction between the chains (e.g. with most nonionic surfactant) or it may increase π as a result of repulsion between the head groups in the case of ionic surfactants.

Various equations of state have been proposed, taking into account the above two effects in order to fit the π-A data. The two-dimensional van der Waals equation of state is probably the most convenient for fitting these adsorption isotherms, i.e.,

$$(\pi)(A - n_2 A_2^0) = n_2 RT,\qquad(4.23)$$

where A_2^0 is the excluded area or co-area of type 2 molecule in the interface and α is a parameter which allows for lateral interaction.

Equation (4.23) leads to the following theoretical adsorption isotherm, using the Gibbs equation:

$$C_2^\alpha = K_1 \left(\frac{\theta}{1-\theta} \right) \exp \left(\frac{\theta}{1-\theta} - \frac{2\alpha\theta}{a_2^0 RT} \right),\qquad(4.24)$$

where θ is the surface coverage ($\theta = \Gamma_2/\Gamma_{2,\max}$), K_1 is a constant that is related to the free energy of adsorption of surfactant molecules at the interface ($K_1 \propto \exp(-\Delta G_{ads}/kT)$) and a_2^0 is the area/molecule.

For a charged surfactant layer, equation (4.21) has to be modified to take into account the electrical contribution from the ionic head groups, i.e.,

$$C_2^\alpha = K_1 \left(\frac{\theta}{1-\theta} \right) \exp \left(\frac{\theta}{1-\theta} \right) \exp \left(\frac{e\psi_0}{kT} \right),\qquad(4.25)$$

where Ψ_0 is the surface potential. Equation (4.47) shows how the electrical potential energy (Ψ_0/kT) of adsorbed surfactant ions affects the surface excess. Assuming that the bulk concentration remains constant, then Ψ_0 increases as θ increases. This means that $[\theta/(1-\theta)] \exp[\theta/(1-\theta)]$ increases less rapidly with C_2, i.e. adsorption is inhibited as a result of ionization.

4.2.3 The Langmuir, Szyszkowski and Frumkin equations

In addition to the Gibbs equation, three other equations have been suggested that relate the surface excess Γ_1, surface or interfacial tension, and equilibrium concentration in the liquid phase C_1. The Langmuir equation [3] relates Γ_1 to C_1 by,

$$\Gamma_1 = \frac{\Gamma_m C_1}{C_1 + a},\qquad(4.26)$$

where Γ_m is the saturation adsorption at monolayer coverage by surfactant molecules. a is a constant that is related to the free energy of adsorption ΔG_{ads}^0,

$$a = 55.3 \exp \left(\frac{\Delta G_{ads}^0}{RT} \right),\qquad(4.27)$$

where R is the gas constant and T is the absolute temperature.

A linear form of the Gibbs equation is

$$\frac{1}{\Gamma_1} = \frac{1}{\Gamma_m} + \frac{a}{\Gamma_m C_1}.\qquad(4.28)$$

Equation (4.28) shows that a plot of $1/\Gamma_1$ versus $1/C_1$ gives a straight line from which Γ_m and a can be calculated from the intercept and slope of the line.

The Szyszkowski equation [4] gives a relationship between the surface pressure π and bulk surfactant concentration C_1; it is a form of an equation of state,

$$\gamma_o - \gamma = \pi = 2.303 RT\Gamma_m \log\left(\frac{C_1}{a} + 1\right). \tag{4.29}$$

The Frumkin equation [5] is another equation of state,

$$\gamma_o - \gamma = \pi = -2.303 RT\Gamma_m \log\left(1 - \frac{\Gamma_1}{\Gamma_m}\right). \tag{4.30}$$

4.3 Interfacial tension measurements

These methods may be classified into two categories: those in which the properties of the meniscus are measured at equilibrium, e.g., pendent drop or sessile drop profile and Wilhelmy plate methods, and those where the measurement is made under non-equilibrium or quasi-equilibrium conditions such as the drop volume (weight) or the de Nouy ring method. The latter methods are faster, although they suffer from the disadvantage of premature rupture and expansion of the interface, causing adsorption depletion. For measurement of low interfacial tensions (< 0.1 mNm^{-1}) the spinning drop technique is applied. Below a brief description of each of these techniques is given.

4.3.1 The Wilhelmy plate method

In this method [6] a thin plate made from glass (e.g., a microscope cover slide) or platinum foil is either detached from the interface (non-equilibrium condition) or it weight measured statically using an accurate microbalance. In the detachment method, the total force F is given by the weight of the plate W and the interfacial tension force,

$$F = W + \gamma p, \tag{4.31}$$

where p is the "contact length" of the plate with the liquid, i.e., the plate perimeter. Provided the contact angle of the liquid is zero, no correction is required for equation (4.31). Thus, the Wilhelmy plate method can be applied in the same manner as du Nouy's technique that is described below.

The static technique may be applied for following the interfacial tension as a function of time (to follow the kinetics of adsorption) till equilibrium is reached. In this case, the plate is suspended from one arm of a microbalance and allowed to penetrate the upper liquid layer (usually the oil) into the aqueous phase to ensure wetting of the plate. The whole vessel is then lowered to bring the plate in the oil phase. At this point the microbalance is adjusted to counteract the weight of the plate (i.e., its weight now becomes zero). The vessel is then raised until the plate touches the interface. The increase in weight ΔW is given by the following equation,

$$\Delta W = \gamma p \cos \theta, \tag{4.32}$$

where θ is the contact angle. If the plate is completely wetted by the lower liquid as it penetrates, $\theta = 0$ and γ may be calculated directly from ΔW. Care should always be taken that the plate is completely wetted by the aqueous solution. For that purpose, a roughened platinum or glass plate is used to ensure a zero contact angle. However, if the oil is denser than water, a hydrophobic plate is used so that when the plate penetrates through the upper aqueous layer and touches the interface it is completely wetted by the oil phase.

4.3.2 The pendent drop method

If a drop of oil is allowed to hang from the end of a capillary that is immersed in the aqueous phase, it will adopt an equilibrium profile shown in Fig. 4.3 that is a unique function of the tube radius, the interfacial tension, its density and the gravitational field.

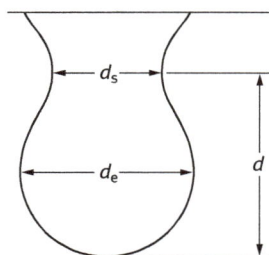

Fig. 4.3. Schematic representation of the profile of a pendent drop.

The interfacial tension is given by the following equation [7]:

$$\gamma = \frac{\Delta \rho g d_e^2}{H}, \tag{4.33}$$

where $\Delta \rho$ is the density difference between the two phases, d_e is the equatorial diameter of the drop (see Fig. 4.3) and H is a function of d_s/d_e, where d_s is the diameter measured at a distance d from the bottom of the drop (see Fig. 4.3). The relationship between H and the experimental values of d_s/d_e has been obtained empirically using pendent drops of water. Accurate values of H have been obtained by Niederhauser and Bartell [8].

4.3.3 The Du Nouy's ring method

Basically one measures the force required to detach a ring or loop of wire from the liquid/liquid interface [9]. As a first approximation, the detachment force is taken to be equal to the interfacial tension γ multiplied by the perimeter of the ring, i.e.,

$$F = W + 4\pi R\gamma,$$ (4.34)

where W is the weight of the ring. Harkins and Jordan [10] introduced a correction factor f (that is a function of meniscus volume V and radius r of the wire) for more accurate calculation of γ from F, i.e.,

$$f = \frac{\gamma}{\gamma_{ideal}} = f\left(\frac{R^3}{V}, \frac{R}{r}\right).$$ (4.35)

Values of the correction factor f were tabulated by Harkins and Jordan [7]. Some theoretical account of f was given by Freud and Freud [11].

When using the du Nuoy method for measurement of γ one must be sure that the ring is kept horizontal during the measurement. Moreover, the ring should be free from any contaminant and this is usually achieved by using a platinum ring that is flamed before use.

4.3.4 The drop volume (weight) method

Here one determines the volume V (or weight W) of a drop of liquid (immersed in the second less dense liquid) which becomes detached from a vertically mounted capillary tip having a circular cross section of radius r. The ideal drop weight W_{ideal} is given by the expression,

$$W_{ideal} = 2\pi r\gamma.$$ (4.36)

In practice, a weight W is obtained which is less than W_{ideal} because a portion of the drop remains attached to the tube tip. Thus, equation (4.36) should include a correction factor, φ, that is a function of the tube radius r and some linear dimension of the drop, i.e., $V^{1/3}$. Thus,

$$W = 2\pi r\gamma\phi\left(\frac{r}{V^{1/3}}\right).$$ (4.37)

Values of $(r/V^{1/3})$ have been tabulated by Harkins and Brown [12]. Lando and Oakley [13] used a quadratic equation to fit the correction function to $(r/V^{1/3})$. A better fit was provided by Wilkinson and Kidwell [14].

4.3.5 The spinning drop method

This method is particularly useful for the measurement of very low interfacial tensions ($< 10^{-1}$ mNm^{-1}) which are particularly important in applications such as spontaneous emulsification and the formation of microemulsions. Such low interfacial tensions may also be reached with emulsions particularly when mixed surfactant films are used. A drop of the less dense liquid A is suspended in a tube containing the second liquid B. On rotating the whole mass (Fig. 4.4) the drop of the liquid moves to the center. With increasing speed of revolution, the drop elongates as the centrifugal force opposes the interfacial tension force that tends to maintain the spherical shape, i.e., that having minimum surface area.

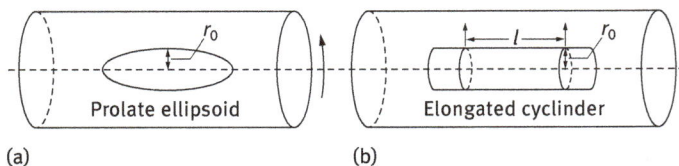

Prolate ellipsoid   Elongated cyclinder

(a)   (b)

Fig. 4.4. Schematic representation of a spinning drop: (a) prolate ellipsoid; (b) elongated cylinder.

An equilibrium shape is reached at any given speed of rotation. At moderate speeds of rotation, the drop approximates to a prolate ellipsoid, whereas at very high speeds of revolution the drop approximates to an elongated cylinder. This is schematically shown in Fig. 4.4.

When the shape of the drop approximates a cylinder (Fig. 4.4b), the interfacial tension is given by the following expression [15],

$$\gamma = \frac{\omega_2 \, \Delta\rho \, r_o^4}{4} \tag{4.38}$$

where ω is the speed of rotation, $\Delta\rho$ is the density difference between the two liquids A and B and r_o is the radius of the elongated cylinder. Equation (4.38) is valid when the length of the elongated cylinder is much larger than r_o.

4.4 Adsorption of surfactants at the solid/liquid interface

As mentioned above, surfactants consist of a small number of units and they are mostly reversibly adsorbed, allowing one to apply thermodynamic treatments. In this case, it is possible to describe the adsorption in terms of the various interaction parameters, namely chain-surface, chain-solvent and surface-solvent. Moreover, the conformation of the surfactant molecules at the interface can be deduced from these simple interaction parameters. However, in some cases these interaction parameters may involve ill-defined forces, such as hydrophobic bonding, solvation forces and chemisorption. In addition, the adsorption of ionic surfactants involves electrostatic forces particularly with polar surfaces containing ionogenic groups. For that reason, the adsorption of ionic and nonionic surfactants will be treated separately. The sur-

faces (substrates) can be also hydrophobic or hydrophilic and these may be treated separately. Thus, four cases can be considered:
1. Adsorption of ionic surfactants on hydrophobic (non-polar) surfaces;
2. Adsorption of ionic surfactants on polar (charged) surfaces;
3. Adsorption of nonionic surfactants on hydrophobic surfaces;
4. Adsorption of nonionic surfactants on polar surfaces.

(1) and (3) are governed by hydrophobic interaction between the alkyl chain and the hydrophobic surface; the charge plays a minor role. (2) and (4) are determined by charge and/or polar interaction.

At the solid/liquid interface one is interested in determining the following parameters:
1. The amount of surfactant adsorbed Γ per unit mass or unit area of the solid adsorbent at a given temperature.
2. The equilibrium concentration of the surfactant C (moles dm^{-3} or mole fraction $x = C/55.51$) in the liquid phase required to produce a given value of Γ at a given temperature.
3. The surfactant concentration at full saturation of the adsorbent Γ_{sat}.
4. The orientation of the adsorbed surfactant ion or molecule that can be obtained from the area occupied by the ion or molecule at full saturation.
5. The effect of adsorption on the properties of the adsorbent (non-polar, polar or charged).

The general equation for calculating the amount of surfactant adsorbed onto a solid adsorbent from a binary solution containing two components (surfactant component 1 and solvent component 2) is given by [16]

$$\frac{n_0 \Delta x_1}{m} = n_1^s x_2 - n_2^s x_1, \tag{4.39}$$

where n_0 is the number of moles of solution before adsorption, $\Delta x_1 = x_{1,0} - x_1$, $x_{1,0}$ is the mole fraction of component 1 before adsorption, x_1 and x_2 are the mole fractions of components 1 and 2 at adsorption equilibrium, m is the mass of adsorbent in grams, n_1^s and n_2^s are the number of components 1 and 2 adsorbed per gram of adsorbent at adsorption equilibrium.

When the liquid phase is a dilute solution of surfactant (component 1) that much more strongly adsorbed onto the solid substrate than the solvent (component 2), then $n_0 \Delta x_1 = \Delta n_1$ where Δn_1 = change in number of moles of component 1 in solution, $n_2^s \approx 0$ and $x_2 \approx 1$. In this case equation (4.39) reduces to

$$n_1^s = \frac{\Delta n_1}{m} = \frac{\Delta C_1 V}{m}, \tag{4.40}$$

where $\Delta C_1 = C_{1,0} - C_1$, $C_{1,0}$ is the molar concentration of component 1 before adsorption, C_1 is the molar concentration of component 1 after adsorption and V is the volume of the liquid phase in liters.

The surface concentration Γ_1 in moles m^{-2} can be calculated from a knowledge of surface area $A(m^2 g^{-1})$,

$$\Gamma_1 = \frac{\Delta C_1 V}{mA}. \tag{4.41}$$

The adsorption isotherm is represented by a plot of Γ_1 versus C_1. In most cases, the adsorption increases gradually with increasing C_1 and a plateau Γ_1^∞ is reached at full coverage corresponding to a surfactant monolayer. The area per surfactant molecule or ion at full saturation can be calculated

$$a_1^s = \frac{10^{18}}{\Gamma_1^\infty N_{av}} nm^2, \tag{4.42}$$

where N_{av} is the Avogadro's number.

4.4.1 Adsorption of ionic surfactants on hydrophobic surfaces

The adsorption of ionic surfactants on hydrophobic surfaces such as carbon black, polymer surfaces and ceramics (silicon carbide or silicon nitride) is governed by hydrophobic interaction between the alkyl chain of the surfactant and the hydrophobic surface. In this case, electrostatic interaction will play a relatively smaller role. However, if the surfactant head group is of the same sign of charge as that on the substrate surface, electrostatic repulsion may oppose adsorption. In contrast, if the head groups are of opposite sign to the surface, adsorption may be enhanced. Since the adsorption depends on the magnitude of the hydrophobic bonding free energy, the amount of surfactant adsorbed increases directly with increasing alkyl chain length in accordance with Traube's rule.

The adsorption of ionic surfactants on hydrophobic surfaces may be represented by the Stern–Langmuir isotherm [17]. Consider a substrate containing N_s sites (mol m^{-2}) on which Γ moles m^{-2} of surfactant ions are adsorbed. The surface coverage θ is (Γ/N_s) and the fraction of uncovered surface is $(1 - \theta)$. The rate of adsorption is proportional to the surfactant concentration expressed in mole fraction, $(C/55.5)$, and the fraction of free surface, $(1 - \theta)$, i.e.

$$\text{Rate of Adsorption} = k_{ads} \left(\frac{C}{55.5} \right) (1 - \theta), \tag{4.43}$$

where k_{ads} is the rate constant for adsorption.

The rate of desorption is proportional to the fraction of surface covered θ,

$$\text{Rate of desorption} = k_{des} \theta. \tag{4.44}$$

At equilibrium, the rate of adsorption is equal to the rate of desorption and the ratio of (k_{ads}/k_{des}) is the equilibrium constant K, i.e.,

$$\frac{\theta}{(1-\theta)} = \frac{C}{55.5} K. \tag{4.45}$$

The equilibrium constant K is related to the standard free energy of adsorption by,

$$-\Delta G^o_{ads} = RT \ln K, \tag{4.46}$$

R is the gas constant and T is the absolute temperature. Equation (4.46) can be written in the form

$$K = \exp\left(-\frac{\Delta G^o_{ads}}{RT}\right). \tag{4.47}$$

Combining equations (4.45) and (4.47),

$$\frac{\theta}{1-\theta} = \frac{C}{55.5} \exp\left(-\frac{\Delta G^o_{ads}}{RT}\right). \tag{4.48}$$

Equation (4.48) applies only at low surface coverage ($\theta < 0.1$) where lateral interaction between the surfactant ions can be neglected.

At high surface coverage ($\theta > 0.1$) one should take the lateral interaction between the chains into account, by introducing a constant A, e.g. using the Frumkin–Fowler–Guggenheim (FFG) equation [17],

$$\frac{\theta}{(1-\theta)} \exp(A\theta) = \frac{C}{55.5} \exp\left(-\frac{\Delta G^o_{ads}}{RT}\right). \tag{4.49}$$

The value of A can be estimated from the maximum slope $(d\theta/\ln C)_{max}$ of the isotherm which occurs at $\theta = 0.5$. Furthermore, at $\theta = 0.5$ substitution of A into equation (4.49) gives the value of ΔG^o_{ads}.

The above treatment using the FFG isotherm has two limitations. Firstly, it is assumed that A is constant and independent of surface coverage. In reality, A could change in sign as well as increase in θ. At low coverages, A would reflect repulsive (electrostatic) interaction between adsorbed surfactant ions. At higher coverage, attractive chain-chain interaction becomes more important. The apparent adsorption energy becomes more favorable at high surface coverage and this could lead to the formation of "hemimicelles". Secondly, electrostatic interactions are strongly affected by the level of supporting electrolyte.

Various authors [18, 19] have used the Stern–Langmuir equation in a simple form to describe the adsorption of surfactant ions on mineral surfaces,

$$\Gamma = 2 \, r \, C \exp\left(-\frac{\Delta G^o_{ads}}{RT}\right). \tag{4.50}$$

Various contributions to the adsorption free energy may be envisaged. To a first approximation, these contributions may be considered to be additive. In the first instance, ΔG_{ads} may be taken to consist of two main contributions, i.e.,

$$\Delta G_{ads} = \Delta G_{elec} + \Delta G_{spec}, \tag{4.51}$$

where ΔG_{elec} accounts for any electrical interactions (coulombic as well as polar) and ΔG_{spec} is a specific adsorption term which contains all contributions to the adsorption free energy that are dependent on the "specific" (non-electrical) nature of the system [20]. Several authors subdivided ΔG_{spec} into supposedly separate independent interactions [20, 21], e.g.

$$\Delta G_{spec} = \Delta G_{cc} + \Delta G_{cs} + \Delta G_{hs} + \ldots\ldots\ldots, \tag{4.52}$$

where ΔG_{cc} is a term that accounts for the cohesive chain-chain interaction between the hydrophobic moieties of the adsorbed ions, ΔG_{cs} is the term for chain/substrate interaction whereas ΔG_{hs} is a term for the head group/substrate interaction. Several other contributions to ΔG_{spec} may be envisaged, e.g. ion-dipole, ion-induced dipole or dipole-induced dipole interactions.

Since there is no rigorous theory that can predict adsorption isotherms, the most suitable method to investigate adsorption of surfactants is to determine the adsorption isotherm. Measurement of surfactant adsorption is fairly straightforward. A known mass m (g) of the particles (substrate) with known specific surface area $A_s (m^2 g^{-1})$ is equilibrated at constant temperature with surfactant solution with initial concentration C_1. The suspension is kept stirred for sufficient time to reach equilibrium. The particles are then removed from the suspension by centrifugation and the equilibrium concentration C_2 is determined using a suitable analytical method. The amount of adsorption Γ (mole m^{-2}) is calculated as follows,

$$\Gamma = \frac{(C_1 - C_2)}{mA_s}. \tag{4.53}$$

The adsorption isotherm is represented by plotting Γ versus C_2. A range of surfactant concentrations should be used to cover the whole adsorption process, i.e. from the initial values low to the plateau values. To obtain accurate results, the solid should have a high surface area (usually $> 1\,m^2$).

Several examples may be quoted from the literature to illustrate the adsorption of surfactant ions on solid surfaces. For a model hydrophobic surface, carbon black has been chosen [22, 23]. Fig. 4.5 shows typical results for the adsorption of sodium dodecyl sulfate (SDS) on two carbon black surfaces, namely Spheron 6 (untreated) and Graphon (graphitized) which also describes the effect of surface treatment.

The adsorption of SDS on untreated Spheron 6 tends to show a maximum that is removed on washing. This suggests the removal of impurities from the carbon black which becomes extractable at high surfactant concentration. The plateau adsorption value is $\sim 2 \times 10^{-6}$ mol m^{-2} ($\sim 2\mu$ mole m^{-2}). This plateau value is reached at ~ 8 m mole dm^{-3} SDS, i.e. close to the cmc of the surfactant in the bulk solution. The area per surfactant ion in this case is $\sim 0.7\,nm^2$. Graphitization (Graphon) removes the hydrophilic ionizable groups (e.g. $-C=O$ or $-COOH$), producing a surface that is more hydrophobic. The same occurs by heating Spheron 6 to 2700°C. This leads to a different adsorption isotherm (Fig. 4.5) showing a step (inflection point) at a surfactant concentration in

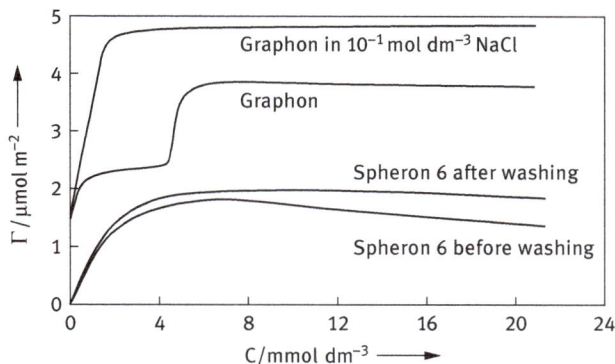

Fig. 4.5. Adsorption isotherms for sodium dodecyl sulfate on carbon substrates.

the region of ~ 6 m mole dm^{-3}. The first plateau value is ~ 2.3 μ mole m^{-2} whereas the second plateau value (that occurs at the cmc of the surfactant) is $\sim \mu$ mole m^{-2}. It is likely in this case that the surfactant ions adopt different orientations at the first and second plateaus. In the first plateau region, a more "flat" orientation (alkyl chains adsorbing parallel to the surface) is obtained, whereas at the second plateau vertical orientation is more favorable, with the polar head groups being directed towards the solution phase. Addition of electrolyte (10^{-1} mole dm^{-3} NaCl) enhances the surfactant adsorption. This increase is due to the reduction of lateral repulsion between the sulfate head groups and this enhances the adsorption.

The adsorption of ionic surfactants on hydrophobic polar surfaces resembles that for carbon black [24, 25]. For example, Saleeb and Kitchener [24] found a similar limiting area for cetyltrimethyl ammonium bromide on Graphon and polystyrene (~ 0.4 nm^2). As with carbon black, the area per molecule depends on the nature and amount of added electrolyte. This can be accounted for in terms of reduction of head group repulsion and/or counterion binding.

Surfactant adsorption close to the cmc may appear Langmuirian, although this does not automatically imply a simple orientation. For example, rearrangement from horizontal to vertical orientation or electrostatic interaction and counterion binding may be masked by simple adsorption isotherms. It is essential, therefore, to combine the adsorption isotherms with other techniques such as microcalorimetry and various spectroscopic methods to obtain a full picture of surfactant adsorption.

4.4.2 Adsorption of ionic surfactants on polar surfaces

The adsorption of ionic surfactants on polar surfaces that contain ionizable groups may show characteristic features due to additional interaction between the head group and substrate and/or possible chain-chain interaction. This is best illus-

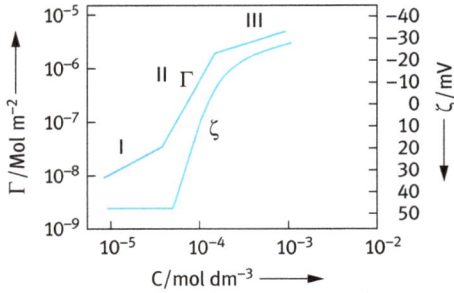

Fig. 4.6. Adsorption isotherm for sodium dodecyl sulfonate on alumina and corresponding zeta (ζ) potential.

trated by the results of adsorption of sodium dodecyl sulfonate (SDSe) on alumina at pH = 7.2 obtained by Fuersetenau [26] and shown in Fig. 4.6. At the pH value, the alumina is positively charged (the isoelectric point of alumina is at pH ~9) and the counterions are Cl$^-$ from the added supporting electrolyte. In Fig. 4.6, the saturation adsorption Γ_1 is plotted versus equilibrium surfactant concentration C_1 in logarithmic scales. The figure also shows the results of zeta potential (ζ) measurements (which are a measure of the magnitude sign of charge on the surface). Both adsorption and zeta potential results show three distinct regions. The first region which shows a gradual increase of adsorption with increasing concentration, with virtually no change in the value of the zeta potential, corresponds to an ion-exchange process [27]. In other words, the surfactant ions simply exchange with the counterions (Cl$^-$) of the supporting electrolyte in the electrical double layer. At a critical surfactant concentration, the desorption increases dramatically with a further increase in surfactant concentration (region II). In this region, the positive zeta potential gradually decrease, reaching a zero value (charge neutralization) after which a negative value is obtained which increases rapidly with increasing surfactant concentration. The rapid increase in region II was explained in terms of "hemimicelle formation" that was originally postulated by Gaudin and Fuerestenau [28]. In other words, at a critical surfactant concentration (to be denoted the cmc of "hemimicelle formation" or better the critical aggregation concentration CAC), the hydrophobic moieties of the adsorbed surfactant chains are "squeezed out" from the aqueous solution by forming two-dimensional aggregates on the adsorbent surface. This is analogous to the process of micellization in bulk solution. However, the CAC is lower than the cmc, indicating that the substrate promotes surfactant aggregation. At a certain surfactant concentration in the hemimicellization process, the isoelectric point is exceeded and, thereafter, the adsorption is hindered by the electrostatic repulsion between the hemimicelles and the micelles and hence the slope of the adsorption isotherm is reduced (region III).

4.4.3 Adsorption of nonionic surfactants

Several types of nonionic surfactants exist, depending on the nature of the polar (hydrophilic) group. The most common type is that based on a poly(oxyethylene) glycol group, i.e. $(CH_2CH_2O)_nOH$ (where n can vary from as little as 2 units to as high as 100 or more units) linked either to an alkyl (C_xH_{2x+1}) or alkyl phenyl $(C_xH_{2x+1}-C_6H_4-)$ group. These surfactants may be abbreviated as C_xE_n or $C_x\phi E_n$ (where C refers to the number of C atoms in the alkyl chain, ϕ denotes C_6H_4 and E denotes ethylene oxide). These ethoxylated surfactants are characterized by a relatively large head group compared to the alkyl chain (when n > 4). However, there are nonionic surfactants with a small head group such as amine oxides $(-N- > 0)$ head group, phosphate oxide $(-P- > 0)$ or sulfinyl-alkanol $(-SO-(CH_2)_n-OH)$. Most adsorption isotherms in the literature are based on the ethoxylated type surfactants.

The adsorption isotherm of nonionic surfactants are in many cases Langmuirian, like those of most other highly surface active solutes adsorbing from dilute solutions, and adsorption is generally reversible. However, several other adsorption types are produced [29] and those are illustrated in Fig. 4.7. The steps in the isotherm may be explained in terms of the various adsorbate-adsorbate, adsorbate-adsorbent and adsorbate-solvent interactions. These orientations are schematically illustrated in Fig. 4.8. In the first stage of adsorption (denoted by I in Figs. 4.7 and 4.8), surfactant-surfactant interaction is negligible (low coverage) and adsorption occurs mainly by van der Waals interaction. On a hydrophobic surface, the interaction is dominated by the hydrophobic portion of the surfactant molecule. This is mostly the case with agrochemicals which have hydrophobic surfaces. However, if the chemical is hydrophilic in nature, the interaction will be dominated by the EO chain. The approach to monolayer saturation with the molecules lying flat is accompanied by a gradual decrease in the slope of the adsorption isotherm (region II in Fig. 4.7). Increase in the size of the surfactant molecule, e.g. increasing the length of the alkyl or EO chain, will decrease adsorption (when expressed in moles per unit area). On the other hand, increasing temperature will increase adsorption as a result of desolvation of the EO chains, thus

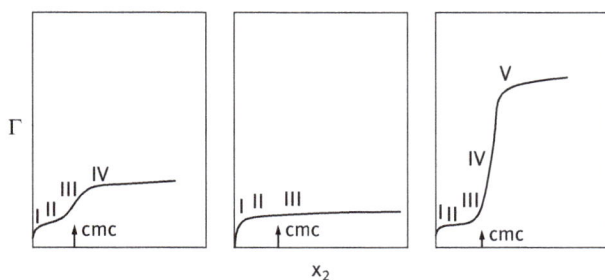

Fig. 4.7. Adsorption isotherms corresponding to the three adsorption sequences shown in Fig. 4.8.

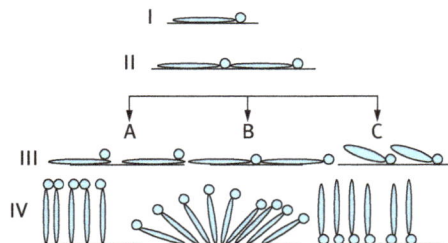

Fig. 4.8. Model for adsorption of nonionic surfactants.

reducing their size. Moreover, increasing temperature reduces the solubility of the nonionic surfactant and this enhances adsorption.

The subsequent stages of adsorption (regions III and IV) are determined by surfactant-surfactant interaction, although surfactant-surface interaction initially determines adsorption beyond stage II. This interaction depends on the nature of the surface and the hydrophilic-lipophilic balance of the surfactant molecule (HLB). For a hydrophobic surface, adsorption occurs via the alkyl group of the surfactant. For a given EO chain, the adsorption will increase with increasing alkyl chain length. On the other hand, for a given alkyl chain length, adsorption increases with a decrease of the PEO chain length.

As the surfactant concentration approaches the cmc, there is a tendency for aggregation of the alkyl groups. This will cause vertical orientation of the surfactant molecules (stage IV). They will compress the head group and for an EO chain this will result in a less coiled, more extended conformation. The larger the surfactant alkyl chain, the greater will be the cohesive forces and hence the smaller the cross-sectional area. This may explain why saturation adsorption increases with increasing alkyl chain length.

The interactions occurring in the adsorption layer during the fourth and subsequent stages of adsorption are similar to those that occur in bulk solution. In this case aggregate units, as shown in Fig. 4.8 V (hemimicelles or micelles) may be formed. This picture was supported by Kleminko et al. [30] who found close agreement between saturation adsorption and adsorption calculated based on the assumption that the surface is covered with close-packed hemimicelles. Kleminco [31] developed a theoretical model for the three stages of adsorption of nonionic surfactants. In the first stage (flat orientation) a modified Langmuir adsorption equation was used. In the second stage of horizontal orientation, the surface concentration increases by an amount that is determined by the displacement of the ethoxy chain by the alkyl group. Finally, in the region of hemimicelle formation, the adsorption can be described by a simple Langmuir equation of the form

$$C_2 K_a^* = \frac{\Gamma_2}{(\Gamma_2^{\text{infinity}} - \Gamma_2)}, \tag{4.54}$$

where Γ_2^∞ is the maximum surface excess, i.e. the surface excess when the surface is covered with close-packed hemimicelles, K_a^* is a constant that is inversely proportional to the cmc and C_2 is the equilibrium concentration.

References

[1]   E. A. Guggenheim, *Thermodynamics*, p. 45, North Holland, Amsterdam, 5th ed. 1967.
[2]   J. W. Gibbs, *Collected Works*, p. 219, Longman, New York, Vol. 1 1928.
[3]   I. Langmuir, *J. Am. Chem. Soc.*, **39**, 1848 (1917).
[4]   B. Szyszkowski, *Z. Phys. Chem.*, **64**, 385 (1908).
[5]   A. Frumkin, *Z. Phys. Chem.*, **116**, 466 (1925).
[6]   L. Wilhelmy, *Ann. Phys.*, **119**, 177 (1863).
[7]   F. Bashforth and J. C. Adams, *An Attempt to Test the Theories of Capillary Action*, University Press, Cambridge, 1883.
[8]   D. O. Nierderhauser and F. E. Bartell, *Report of Progress, Fundamental Research on Occurence of Petroleum*, Publication of the American Petroleum Institute, Lord Baltimore Press, Baltimore, Md., 1950, p. 114.
[9]   P. L. Du Nouy, *J. Gen. Physiol.*, **1**, 521 (1919).
[10]   W. D. Harkins and H.F. Jordan, *J. Amer. Chem. Soc.*, **52**, 1715 (1930).
[11]   B. B. Freud and H. Z. Freud, *J. Amer. Chem. Soc.*, **52**, 1772 (1930).
[12]   W. D. Harkins, and F. E. Brown, *J. Amer. Chem. Soc.*, **41**, 499 (1919).
[13]   J. L. Lando and H. T. Oakley, *J. Colloid Interface Sci.*, **25**, 526 (1967).
[14]   M. C. Wilkinson and R. L. Kidwell, *J. Colloid Interface Sci.*, **35**, 114 (1971).
[15]   B. Vonnegut, *New Sci. Intrum.*, **13**, 6 (1942).
[16]   R. Aveyard and D. A. Haydon, *An Introduction to the Principles of Surface Chemistry*, Cambridge University Press, Cambridge, 1973.
[17]   D. B. Hough and H. M. Randall, in: *Adsorption from Solution at the Solid/Liquid Interface*, G. D. Parfitt and C. H. Rochester (eds.), p. 247, Academic Press, London, 1983.
[18]   D. W. Fuerstenau and T. W. Healy, in: *Adsorptive Bubble Separation Techniques*, R. Lemlich (ed.), p. 91, Academic Press, London, 1972.
[19]   P. Somasundaran and E. D. Goddard, *Modern Aspects Electrochem.*, **13**, 207 (1979).
[20]   T. W. Healy, *J. Macromol. Sci. Chem.*, **118**, 603 (1974).
[21]   P. Somasundaran and H.S. Hannah, in: *Improved Oil Recovery by Surfactant and Polymer Flooding*, D. O. Shah and R. S. Schechter (eds.), p. 205, Academic Press, London, 1979.
[22]   F. G. Greenwood, G. D. Parfitt, N. H. Picton and D. G. Wharton, *Adv. Chem. Ser.*, **79**, 135 (1968).
[23]   R. E. Day, F. G. Greenwood and G. D. Parfitt, *4th Int. Congress of Surface Active Substances*, **18**, 1005 (1967).
[24]   F. Z. Saleeb and J. A. Kitchener, *J. Chem. Soc.*, 911 (1965).
[25]   P. Conner and R. H. Ottewill, *J. Colloid Interface Sci.*, **37**, 642 (1971).
[26]   D. W. Fuerestenau, in: *The Chemistry of Biosurfaces*, M. L. Hair (ed.), p. 91, Marcel Dekker, New York, 1971.
[27]   T. Wakamatsu and D. W. Fuerstenau, *Adv. Che. Ser.*, **71**, 161, (1968).
[28]   A. M. Gaudin and D. W. Fuertsenau, *Trans. AIME*, **202**, 958 (1955).
[29]   J. S. Clunie and B. T. Ingram, in: *Adsorption from Solution at the Solid/Liquid Interface*, G. D. Parfitt and C. H. Rochester (eds.), p. 105, Academic Press, London, 1983.
[30]   N. A. Klimenko, Tryasorukova and Permilouskayan, *Kolloid. Zh.*, **36**, 678 (1974).
[31]   N. A. Kleminko, *Kolloid Zh.*, **40**, 1105 (1978); **41**, 78 (1979).

5 Surfactants as emulsifiers

5.1 Introduction

Emulsions are a class of disperse systems consisting of two immiscible liquids [1–3]. The liquid droplets (the disperse phase) are dispersed in a liquid medium (the continuous phase). Several classes may be distinguished: Oil-in-Water (O/W); Water-in-Oil (W/O); Oil-in-oil (O/O). The latter class may be exemplified by an emulsion consisting of a polar oil (e.g. propylene glycol) dispersed in a non-polar oil (paraffinic oil) and vice versa. To disperse two immiscible liquids one needs a third component, namely the emulsifier. The choice of the emulsifier is crucial in formation of the emulsion and its long-term stability [1–3].

Emulsions may be classified according to the nature of the emulsifier or the structure of the system. This is illustrated in Table 5.1.

Table 5.1. Classification of emulsion types.

Nature of emulsifier	Structure of the system
Simple molecules and ions	Nature of internal and external phase:
Nonionic surfactants	O/W, W/O
Surfactant mixtures	Micellar emulsions (microemulsions)
Ionic surfactants	Macroemulsions
Nonionic polymers	Bilayer droplets
Polyelectrolytes	Double and multiple emulsions
Mixed polymers and surfactants	Mixed emulsions Liquid crystalline phases
Solid particles	

5.1.1 Nature of the emulsifier

The simplest type is ions such as OH- which can be specifically adsorbed on the emulsion droplet thus producing a charge. An electrical double layer can be produced which provides electrostatic repulsion. This has been demonstrated with very dilute O/W emulsions by removing any acidity. Clearly that process is not practical. The most effective emulsifiers are nonionic surfactants which can be used to emulsify oil in water or water in oil. In addition they can stabilize the emulsion against flocculation and coalescence. Ionic surfactants such as sodium dodecyl sulfate can also be used as emulsifiers (for O/W) but the system is sensitive to the presence of electrolytes. Surfactant mixtures, e.g. ionic and nonionic or mixtures of nonionic surfactants can be more effective in emulsification and stabilization of the emulsion. Nonionic polymers, sometimes referred to as polymeric surfactants, e.g. Pluronics that are A–B–A block copolymers (with A being polyethylene oxide and B being polypropylene oxide), are more effective in stabilization of the emulsion but they may suffer from the difficulty of emulsification (to produce small droplets) unless high energy is applied for the process. Polyelectrolytes such as poly(methacrylic acid) can also be applied as emulsifiers. Mixtures of polymers and surfactants are ideal in achieving ease of emulsification and stabilization of the emulsion. Lamellar liquid crystalline phases that can be produced using surfactant mixtures are very effective in emulsion stabilization. Solid particles that can accumulate at the O/W interface can also be used for emulsion stabilization. These are referred to as Pickering emulsions, whereby particles are made partially wetted by the oil phase and partially wetted by the aqueous phase.

5.1.2 Structure of the system

1. O/W and W/O macroemulsions. These usually have a size range of 0.1–5 µm with an average of 1–2 µm.
2. Nanoemulsions: these usually have a size range 20–100 nm. Like macroemulsions they are only kinetically stable.
3. Micellar emulsions or microemulsions: these usually have the size range 5–50 nm. They are thermodynamically stable.
4. Double and multiple emulsions: these are emulsions-of-emulsions, W/O/W and O/W/O systems.
5. Mixed emulsions: these are systems consisting of two different disperse droplets that do not mix in a continuous medium. The present chapter will only deal with macroemulsions.

Several breakdown processes may occur on storage depending on: particle size distribution and density difference between the droplets and the medium; magnitude of the attractive versus repulsive forces which determine flocculation; solubility of the disperse droplets and the particle size distribution which determines Ostwald ripening; stability of the liquid film between the droplets that determines coalescence; phase inversion.

5.1.3 Breakdown processes in emulsions

The various breakdown processes are illustrated in Fig. 5.1. The physical phenomena involved in each breakdown process are not simple and it requires analysis of the various surface forces involved. In addition, the above processes may take place simultaneously rather than consecutively and this complicates the analysis. Model emulsions, with monodisperse droplets cannot be easily produced and hence any theoretical treatment must take into account the effect of droplet size distribution. Theories

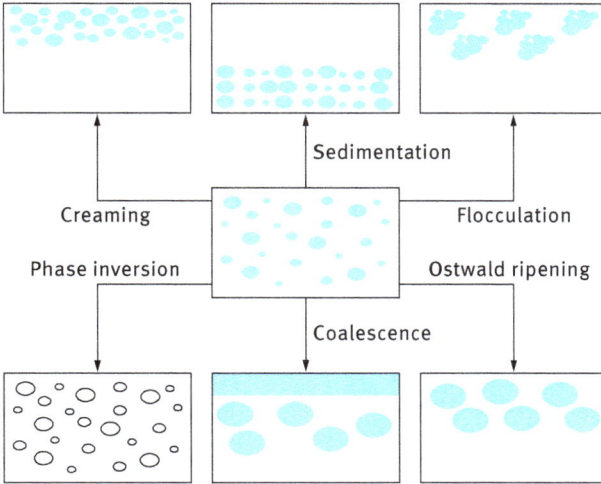

Fig. 5.1. Schematic representation of the various breakdown processes in emulsions.

that take into account the polydispersity of the system are complex and in many cases only numerical solutions are possible. In addition, measurement of surfactant and polymer adsorption in an emulsion is not easy and one has to extract such information from measurements at a planer interface.

Below a summary of each of the above breakdown processes is given and details of each process and methods of its prevention are given separate sections.

5.1.3.1 Creaming and sedimentation

This process results from external forces, usually gravitational or centrifugal. When such forces exceed the thermal motion of the droplets (Brownian motion), a concentration gradient builds up in the system with the larger droplets moving faster to the top (if their density is lower than that of the medium) or to the bottom (if their density is larger than that of the medium) of the container. In the limiting cases, the droplets may form a close-packed (random or ordered) array at the top or bottom of the system with the remainder of the volume occupied by the continuous liquid phase.

5.1.3.2 Flocculation

This process refers to aggregation of the droplets (without any change in primary droplet size) into larger units. It is the result of the van der Waals attraction which is universal with all disperse systems. Flocculation occurs when there is not sufficient repulsion to keep the droplets apart to distances where the van der Waals attraction is weak. Flocculation may be "strong" or "weak", depending on the magnitude of the attractive energy involved.

5.1.3.3 Ostwald ripening (disproportionation)

This results from the finite solubility of the liquid phases. Liquids which are referred to as being immiscible often have mutual solubilities which are not negligible. With emulsions which are usually polydisperse, the smaller droplets will have larger solubility when compared with the larger ones (due to curvature effects). With time, the smaller droplets disappear and their molecules diffuse to the bulk and become deposited on the larger droplets. With time the droplet size distribution shifts to larger values.

5.1.3.4 Coalescence

This refers to the process of thinning and disruption of the liquid film between the droplets with the result of fusion of two or more droplets into larger ones. The limiting case for coalescence is the complete separation of the emulsion into two distinct liquid phases. The driving force for coalescence is the surface or film fluctuation which results in the close approach of the droplets where the van der Waals forces are strong, thus preventing their separation.

5.1.3.5 Phase inversion

This refers to the process whereby there will be an exchange between the disperse phase and the medium. For example an O/W emulsion may with time or change of conditions invert to a W/O emulsion. In many cases, phase inversion passes through a transition state whereby multiple emulsions are produced.

5.1.3.6 Industrial applications of emulsions

Several industrial systems consist of emulsions of which the following are worth mentioning. Food emulsions, e.g. mayonnaise, salad creams, deserts, beverages, etc.; personal care and cosmetics, e.g. hand creams, lotions, hair sprays, sunscreens, etc.; agrochemicals, e.g. self-emulsifiable oils which produce emulsions on dilution with water, emulsion concentrates (EWs) and crop oil sprays; pharmaceuticals, e.g. anesthetics of O/W emulsions, lipid emulsions, double and multiple emulsions, etc.; Paints, e.g. emulsions of alkyd resins, latex emulsions, etc.; dry cleaning formulations: these may contain water droplets emulsified in the dry cleaning oil which is necessary to remove soils and clays; bitumen emulsions: these are emulsions prepared stable in the containers but when applied to the road chippings they must coalesce to form a uniform film of bitumen; emulsions in the oil industry: many crude oils contain water droplets (for example, North Sea oil) and these must be removed by coalescence followed by separation; oil slick dispersions: the oil spilled from tankers must be emulsified and then separated; emulsification of unwanted oil: this is an important process for pollution control.

The above importance of emulsions in industry justifies a great deal of basic research to understand the origin of instability and methods to prevent their breakdown. Unfortunately, fundamental research on emulsions is not easy since model systems (e.g. with monodisperse droplets) are difficult to produce. In many cases, theories on emulsion stability are not exact and semi-empirical approaches are used.

5.2 Physical chemistry of emulsion systems

5.2.1 The interface (Gibbs dividing line)

An interface between two bulk phases, e.g. liquid and air (or liquid/vapor) or two immiscible liquids (oil/water) may be defined provided a dividing line is introduced (Fig. 5.2). The interfacial region is not a layer that is one molecule thick – it is a region with thickness δ with properties different from the two bulk phases α and β.

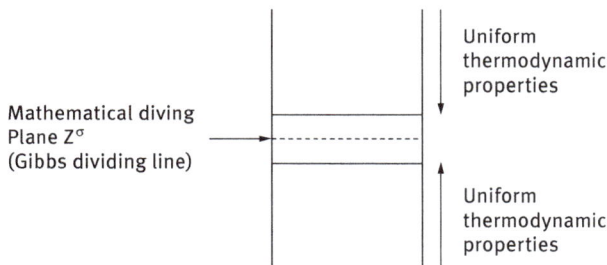

Fig. 5.2. The Gibbs dividing line.

Using the Gibbs model, it is possible to obtain a definition of the surface or interfacial tension γ.

The surface free energy dG^σ is made of three components: An entropy term $S^\sigma dT$; an interfacial energy term $Ad\gamma$; a composition term $\Sigma n_i d\mu_i$ (n_i is the number of moles of component i with chemical potential μ_i). The Gibbs–Deuhem equation is

$$dG^\sigma = -S^\sigma dT + Ad\gamma + \sum n_i d\mu_i . \tag{5.1}$$

At constant temperature and composition,

$$dG^\sigma = Ad\gamma$$

$$\gamma = \left(\frac{\partial G^\sigma}{\partial A}\right)_{T,n_i} . \tag{5.2}$$

For a stable interface γ is positive, i.e. if the interfacial area increases, G^σ increases. Note that γ is energy per unit area (mJm^{-2}) which is dimensionally equivalent to force per unit length (mNm^{-1}), the unit usually used to define surface or interfacial tension.

For a curved interface, one should consider the effect of the radius of curvature. Fortunately, γ for a curved interface is estimated to be very close to that of a planer surface, unless the droplets are very small (< 10 nm). Curved interfaces produce some other important physical phenomena which affect emulsion properties, e.g. the Laplace pressure Δp which is determined by the radii of curvature of the droplets,

$$\Delta p = \gamma \left(\frac{1}{r_1} + \frac{1}{r_2} \right) \tag{5.3}$$

where r_1 and r_2 are the two principal radii of curvature.

For a perfectly spherical droplet $r_1 = r_2 = r$ and

$$\Delta p = \frac{2\gamma}{r} \tag{5.4}$$

For a hydrocarbon droplet with radius 100 nm, and $\gamma = 50$ mNm^{-1}, $\Delta p \sim 10^6$ Pa (10 atm.)

5.2.2 Thermodynamics of emulsion formation and breakdown

Consider a system in which an oil is represented by a large drop 2 of area A_1 immersed in a liquid 2, which is now subdivided into a large number of smaller droplets with total area A_2 ($A_2 \gg A_1$) as shown in Fig. 5.3. The interfacial tension γ_{12} is the same for the large and smaller droplets since the latter are generally in the region of 0.1 to few μm.

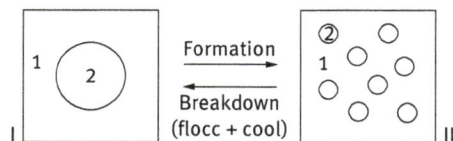

Fig. 5.3. Schematic representation of emulsion formation and breakdown.

The change in free energy in going from state I to state II is made from two contributions: A surface energy term (that is positive) that is equal to $\Delta A\, \gamma_{12}$ (where $\Delta A = A_2 - A_1$), and an entropy of dispersions term which is also positive (since producing a large number of droplets is accompanied by an increase in configurational entropy) which is equal to $T\Delta S^{conf}$.

From the second law of thermodynamics,

$$\Delta G^{form} = \Delta A\gamma_{12} - T\Delta S^{conf}. \tag{5.5}$$

In most cases $\Delta A\gamma_{12} \gg -T\Delta S^{conf}$, which means that ΔG^{form} is positive, i.e. the formation of emulsions is non-spontaneous and the system is thermodynamically unsta-

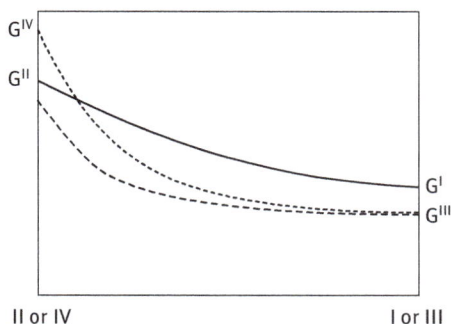

Fig. 5.4. Free energy path in Emulsion breakdown: Flocc. + coal. (black line); Flocc. + coal. + Sed. (dashed line); Flocc. + coal. + sed. + Ostwald ripening (dotted line).

ble. In the absence of any stabilization mechanism, the emulsion will break by flocculation, coalescence, Ostwald ripening or a combination of all these processes. This is illustrated in Fig. 5.4 which shows several paths for emulsion breakdown processes.

In the presence of a stabilizer (surfactant and/or polymer), an energy barrier is created between the droplets and therefore the reversal from state II to state I becomes non-continuous as a result of the presence of these energy barriers. This is illustrated in Fig. 5.5. In the presence of the above energy barriers, the system becomes kinetically stable.

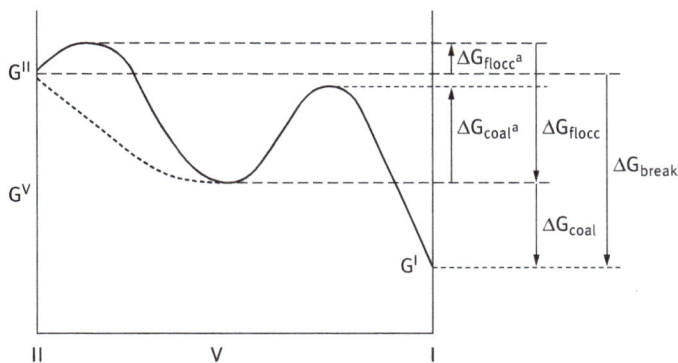

Fig. 5.5. Schematic representation of free energy path for breakdown (flocculation and coalescence) for systems containing an energy barrier.

5.2.3 Interaction energies (forces) between emulsion droplets and their combinations

Generally speaking, there are three main interaction energies (forces) between emulsion droplets and these are discussed below.

5.2.3.1 Van der Waals attraction

The van der Waals attractions between atoms or molecules are of three different types: dipole-dipole (Keesom), dipole-induced dipole (Debye) and dispersion (London) interactions. The Keesom and Debye attraction forces are vectors and although dipole-dipole or dipole-induced dipole attraction is large they tend to cancel due to the different orientations of the dipoles. Thus, the most important are the London dispersion interactions which arise from charge fluctuations. With atoms or molecules consisting of a nucleus and electrons that are continuously rotating around the nucleus, a temporary dipole is created as a result of charge fluctuations. This temporary dipole induces another dipole in the adjacent atom or molecule. The interaction energy between two atoms or molecules G_a is short range and is inversely proportional to the sixth power of the separation distance r between the atoms or molecules,

$$G_a = -\frac{\beta}{r^6} \tag{5.6}$$

where β is the London dispersion constant that is determined by the polarizability of the atom or molecule.

Hamaker [4] suggested that the London dispersion interactions between atoms or molecules in macroscopic bodies (such as emulsion droplets) can be added resulting in strong van der Waals attraction, particularly at close distances of separation between the droplets. For two droplets with equal radii R, at a separation distance h, the van der Waals attraction G_A is given by the following equation (due to Hamaker),

$$G_A = -\frac{AR}{12h}, \tag{5.7}$$

where A is the effective Hamaker constant,

$$A = \left(A_{11}^{1/2} - A_{22}^{1/2}\right)^2, \tag{5.8}$$

where A_{11} and A_{22} are the Hamaker constants of droplets and dispersion medium respectively.

The Hamaker constant of any material depends on the number of atoms or molecules per unit volume q and the London dispersion constant β,

$$A = \pi^2 q^2 \beta. \tag{5.9}$$

G_A increases very rapidly with decrease of h (at close approach). This is illustrated in Fig. 5.6 which shows the van der Waals energy-distance curve for two emulsion droplets with separation distance h.

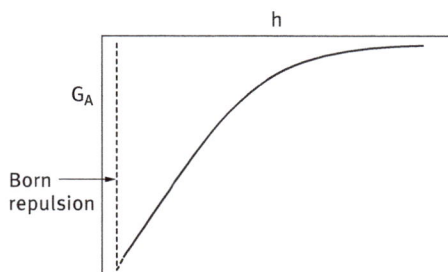

Fig. 5.6. Variation of the van der Waals attraction energy with separation distance.

In the absence of any repulsion, flocculation is very fast producing large clusters. To counteract the van der Waals attraction, it is necessary to create a repulsive force. Two main types of repulsion can be distinguished depending on the nature of the emulsifier used: electrostatic (due to the creation of double layers) and steric (due to the presence of adsorbed surfactant or polymer layers).

5.2.3.2 Electrostatic repulsion

This can be produced by adsorption of an ionic surfactant as is illustrated in Fig. 5.7 which shows a schematic picture of the structure of the double layer according to Gouy–Chapman and Stern pictures [3]. The surface potential ψ_0 decreases linearly to ψ_d (Stern or zeta potential) and then exponentially with increase of distance x. The double layer extension depends on electrolyte concentration and valency (the lower the electrolyte concentration and the lower the valency the more extended the double layer is).

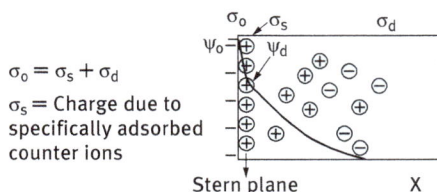

$$\sigma_0 = \sigma_s + \sigma_d$$

σ_s = Charge due to specifically adsorbed counter ions

Fig. 5.7. Schematic representation of double layers produced by adsorption of an ionic surfactant.

When charged colloidal particles in a dispersion approach each other such that the double layer begins to overlap (particle separation becomes less than twice the double layer extension), repulsion occurs. The individual double layers can no longer develop unrestrictedly, since the limited space does not allow complete potential decay [3, 4]. This is illustrated in Fig. 5.8 for two flat plates which clearly shows that when the separation distance h between the emulsion droplets become smaller than twice the double layer extension, the potential at the mid-plane between the surfaces is not

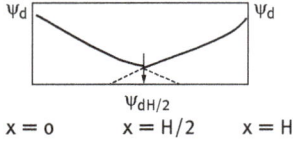

Fig. 5.8. Schematic representation of double layer overlap.

equal to zero (which would be the case when h is larger than twice the double layer extension).

The repulsive interaction G_{el} is given by the following expression,

$$G_{el} = 2\pi R \epsilon_r \epsilon_o \psi_o^2 \ln\left[1 + \exp(-\kappa h)\right] \qquad (5.10)$$

where ϵ_r is the relative permittivity and ϵ_o is the permittivity of free space.

κ is the Debye–Huckel parameter; $1/\kappa$ is the extension of the double layer (double layer thickness) that is given by the expression,

$$\left(\frac{1}{\kappa}\right) = \left(\frac{\epsilon_r \epsilon_o kT}{2n_o Z_i^2 e^2}\right) \qquad (5.11)$$

where k is the Boltzmann constant, T is the absolute temperature, n_o is the number of ions per unit volume of each type present in bulk solution, Z_i is the valency of the ions and e is the electronic charge.

Values of $(1/\kappa)$ at various $1:1$ electrolyte concentrations are given below:

$C/mol dm^{-3}$	10^{-5}	10^{-4}	10^{-3}	10^{-2}	10^{-1}
$(1/\kappa)/nm$	100	33	10	3.3	1

The double layer extension decreases with increasing electrolyte concentration. This means that the repulsion decreases with increasing electrolyte concentration as is illustrated in Fig. 5.9.

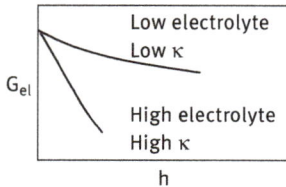

Fig. 5.9. Variation of G_{el} with h at low and high electrolyte concentrations.

Combination of van der Waals attraction and double layer repulsion results in the well-known theory of colloid stability due to Deryaguin, Landau, Verwey and Overbeek (DLVO theory) [5, 6].

$$G_T = G_{el} + G_A. \tag{5.12}$$

A schematic representation of the force (energy) distance curve according to the DLVO theory is given in Fig. 5.10.

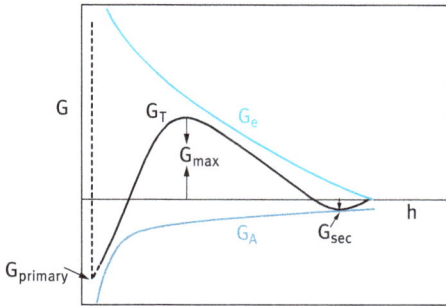

Fig. 5.10. Total energy-distance curve according to the DLVO theory.

The above presentation is for a system at low electrolyte concentration. At large h, attraction prevails resulting in a shallow minimum (G_{sec}) of the order of few kT units. At very short h, $V_A \gg G_{el}$, resulting in a deep primary minimum (several hundred kT units). At intermediate h, $G_{el} > G_A$ resulting in a maximum (energy barrier) whose height depends on ψ_o (or ζ) and electrolyte concentration and valency – the energy maximum is usually kept >25 kT units. The energy maximum prevents close approach of the droplets and flocculation into the primary minimum is prevented. The higher the value of ψ_o and the lower the electrolyte concentration and valency, the higher the energy maximum. At intermediate electrolyte concentrations, weak flocculation into the secondary minimum may occur.

5.2.3.3 Steric repulsion
This is produced by using nonionic surfactants or polymers, e.g. alcohol ethoxylates, or A–B–A block copolymers PEO–PPO–PEO (where PEO refers to polyethylene oxide and PPO refers to polypropylene oxide), as is illustrated in Fig. 5.11.

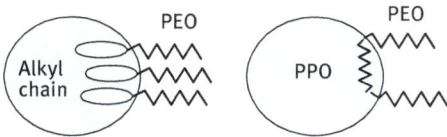

Fig. 5.11. Schematic representation of adsorbed layers.

The "thick" hydrophilic chains (PEO in water) produce repulsion as a result of two main effects [7]:

(a) Unfavorable mixing of the PEO chains, when these are in good solvent conditions (moderate electrolyte and low temperatures). This is referred to as the osmotic or mixing free energy of interaction that is given by the expression,

$$\frac{G_{mix}}{kT} = \left(\frac{4\pi}{V_1}\right)\phi_2^2 N_{av}\left(\frac{1}{2} - \chi\right)\left(\delta - \frac{h}{2}\right)^2\left(3R + 2\delta + \frac{h}{2}\right) \tag{5.13}$$

V_1 is the molar volume of the solvent, ϕ_2 is the volume fraction of the polymer chain with a thickness δ and χ is the Flory–Huggins interaction parameter.

When $\chi < 0.5$, G_{mix} is positive and the interaction is repulsive. When $\chi > 0.5$, G_{mix} is negative and the interaction is attractive. When $\chi = 0.5$, $G_{mix} = 0$ and this is referred to as the θ-condition.

(b) Entropic, volume restriction or elastic interaction, G_{el}. This results from the loss in configurational entropy of the chains on significant overlap. Entropy loss is unfavorable and, therefore, G_{el} is always positive.

Combination of G_{mix}, G_{el} with G_A gives the total energy of interaction G_T (Theory of steric stabilization),

$$G_T = G_{mix} + G_{el} + G_A. \tag{5.14}$$

A schematic representation of the variation of G_{mix}, G_{el} and G_A with h is given in Fig. 5.11. G_{mix} increases very sharply with decrease of h when the latter becomes less than 2δ. G_{el} increase very sharply with decrease of h when the latter becomes smaller than δ. G_T increases very sharply with decrease of h when the latter becomes less than 2δ.

Fig. 5.12 shows that there is only one minimum (G_{min}) whose depth depends on R, δ and A. At a given droplet size and Hamaker constant, the larger the adsorbed layer thickness, the smaller the depth of the minimum. If G_{min} is made sufficiently small (large δ and small R), one may approach thermodynamic stability. This is illustrated in Fig. 5.13 which shows the energy-distance curves as a function of δ/R. The larger the value of δ/R, the smaller the value of G_{min}. In this case the system may approach thermodynamic stability as is the case with nanodispersions

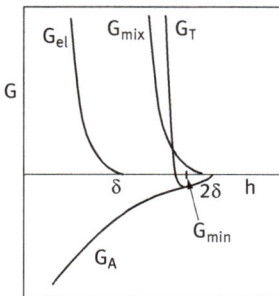

Fig. 5.12. Schematic representation of the energy-distance curve for a sterically stabilized emulsion.

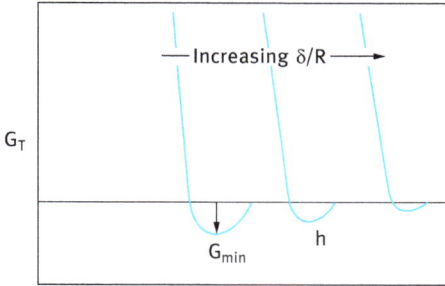

Fig. 5.13. Variation of G_T with h at various δ/R values.

5.3 Mechanism of emulsification

As mentioned before, to prepare an emulsion oil, water, surfactant and energy are needed. This can be considered from a consideration of the energy required to expand the interface, $\Delta A \gamma$ (where ΔA is the increase in interfacial area when the bulk oil with area A_1 produces a large number of droplets with area A_2; $A_2 \gg A_1$; γ is the interfacial tension). Since γ is positive, the energy to expand the interface is large and positive; this energy term cannot be compensated by the small entropy of dispersion $T\Delta S$ (which is also positive) and the total free energy of formation of an emulsion, ΔG given by equation (5.6) is positive. Thus, emulsion formation is non-spontaneous and energy is required to produce the droplets.

The formation of large droplets (few μm) as is the case for macroemulsions is fairly easy and hence high speed stirrers such as the Ultraturrax or Silverson Mixer are sufficient to produce the emulsion. In contrast, the formation of small drops (submicron as is the case with nanoemulsions) is difficult and this requires a large amount of surfactant and/or energy. The high energy required for formation of nanoemulsions can be understood from a consideration of the Laplace pressure Δp (the difference in pressure between inside and outside the droplet) as given by equations (5.3) and (5.4)

To break up a drop into smaller ones, it must be strongly deformed and this deformation increases Δp. Since the stress is generally transmitted by the surrounding liquid via agitation, higher stresses need more vigorous agitation, and hence more energy is needed to produce smaller drops.

Surfactants play major roles in the formation of emulsions: by lowering the interfacial tension, p is reduced and hence the stress needed to break up a drop is reduced. Surfactants also prevent coalescence of newly-formed drops. To describe emulsion formation one has to consider two main factors: hydrodynamics and interfacial science. In hydrodynamics one has to consider the type of flow: laminar flow and turbulent flow. This depends on the Raynolds number as will be discussed later.

To assess emulsion formation, one usually measures the droplet size distribution using for example laser diffraction techniques. A useful average diameter d is,

$$d_{nm} = \left(\frac{S_m}{S_n} \right)^{1/(n-m)}.$$

(5.15)

In most cases d_{32} (the volume/surface average or Sauter mean) is used. The width of the size distribution can be given as the variation coefficient c_m which is the standard deviation of the distribution weighted with d^m divided by the corresponding average d. Generally C_2 will be used which corresponds to d_{32}.

An alternative way to describe the emulsion quality is to use the specific surface area A (surface area of all emulsion droplets per unit volume of emulsion),

$$A = \pi s^2 = \frac{6\phi}{d_{32}}.$$

(5.16)

5.3.1 Methods of emulsification

Several procedures may be applied for emulsion preparation, these range from simple pipe flow (low agitation energy L), static mixers and general stirrers (low to medium energy, L–M), high speed mixers such as the Ultraturrex (M), colloid mills and high pressure homogenizers (high energy, H), ultrasound generators (M–H). The method of preparation can be continuous (C) or batch-wise (B): pipe flow and static mixers: C; stirrers and Ultraturrax: B,C; colloid mill and high pressure homogenizers: C; ultrasound: B,C.

In all methods, there is liquid flow: unbounded and strongly confined flow. In the unbounded flow any droplets is surrounded by a large amount of flowing liquid (the confining walls of the apparatus are far away from most of the droplets). The forces can be frictional (mostly viscous) or inertial. Viscous forces cause shear stresses to act on the interface between the droplets and the continuous phase (primarily in the direction of the interface). The shear stresses can be generated by laminar flow (LV) or turbulent flow (TV); this depends on the Raynolds number R_e,

$$R_e = \frac{vl\rho}{\eta},$$

(5.17)

where v is the linear liquid velocity, ρ is the liquid density and η is its viscosity. l is a characteristic length that is given by the diameter of flow through a cylindrical tube and by twice the slit width in a narrow slit.

For laminar flow $R_e \lesssim 1000$, whereas for turbulent flow $R_e \gtrsim 2000$. Thus whether the regime is linear or turbulent depends on the scale of the apparatus, the flow rate and the liquid viscosity [9–12]. If the turbulent eddies are much larger than the droplets, they exert shear stresses on the droplets. If the turbulent eddies are much smaller than the droplets, inertial forces will cause disruption (TI).

In bounded flow other relations hold. If the smallest dimension of the part of the apparatus in which the droplets are disrupted (say a slit) is comparable to droplet size, other relations hold (the flow is always laminar). A different regime prevails if the droplets are directly injected through a narrow capillary into the continuous phase (injection regime), i.e. membrane emulsification.

Within each regime, an essential variable is the intensity of the forces acting; the viscous stress during laminar flow $\sigma_{viscous}$ is given by

$$\sigma_{viscous} = \eta G, \tag{5.18}$$

where G is the velocity gradient.

The intensity in turbulent flow is expressed by the power density ϵ (the amount of energy dissipated per unit volume per unit time); for laminar flow,

$$\epsilon = \eta G^2. \tag{5.19}$$

The most important regimes are: Laminar/Viscous (LV); Turbulent/Viscous (TV); Turbulent/Inertial (TI). For water as the continuous phase, the regime is always TI. For higher viscosity of the continuous phase ($\eta_C = 0.1$ Pas), the regime is TV. For still higher viscosity or a small apparatus (small l), the regime is LV. For very small apparatus (as is the case with most laboratory homogenizers), the regime is nearly always LV.

For the above regimes, a semi-quantitative theory is available that can give the timescale and magnitude of the local stress σ_{ext}, the droplet diameter d, timescale of droplets deformation τ_{def}, timescale of surfactant adsorption, τ_{ads} and mutual collision of droplets.

An important parameter that describes droplet deformation is the Weber number W_e (which gives the ratio of the external stress over the Laplace pressure),

$$W_e = \frac{G\eta_C R}{2\gamma}. \tag{5.20}$$

The viscosity of the oil plays an important role in the break-up of droplets; the higher the viscosity, the longer it will take to deform a drop. The deformation time τ_{def} is given by the ratio of oil viscosity to the external stress acting on the drop,

$$\tau_{def} = \frac{\eta_D}{\sigma_{ext}}. \tag{5.21}$$

The viscosity of the continuous phase η_C plays an important role in some regimes: for a turbulent inertial regime, η_C has no effect on droplets size. For a turbulent viscous regime, larger ηC leads to smaller droplets. For a laminar viscous regime, the effect is even stronger.

5.3.2 Role of surfactants in emulsion formation

Surfactants lower the interfacial tension γ and this causes a reduction in droplet size. The latter decreases with decreasing γ. For laminar flow the droplet diameter is proportional to γ; for the turbulent inertial regime, the droplet diameter is proportional to $\gamma^{3/5}$.

The effect of reducing γ on the droplet size is illustrated in Fig. 5.14 which shows a plot of the droplet surface area A and mean drop size d_{32} as a function of surfactant concentration m for various systems.

The amount of surfactant required to produce the smallest drop size will depend on its activity a (concentration) in the bulk which determines the reduction in γ, as given by the Gibbs adsorption equation,

$$-d\gamma = RT\Gamma\, d\ln a,$$

(5.22)

where R is the gas constant, T is the absolute temperature and Γ is the surface excess (number of moles adsorbed per unit area of the interface).

Γ increases with increasing surfactant concentration and eventually it reaches a plateau value (saturation adsorption). This is illustrated in Fig. 5.15 for various emulsifiers.

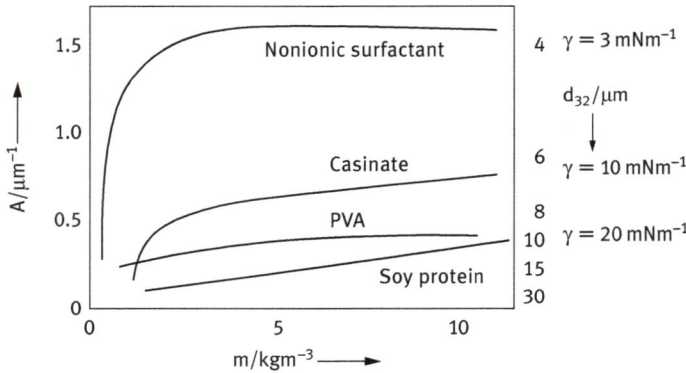

Fig. 5.14. Variation of A and d_{32} with m for various surfactant systems.

The value of γ obtained depends on the nature of the oil and surfactant used; small molecules such as nonionic surfactants lower γ more than polymeric surfactants such as PVA.

Another important role of the surfactant is its effect on the interfacial dilational modulus ϵ,

$$\epsilon = \frac{d\gamma}{d\ln A}.$$

(5.23)

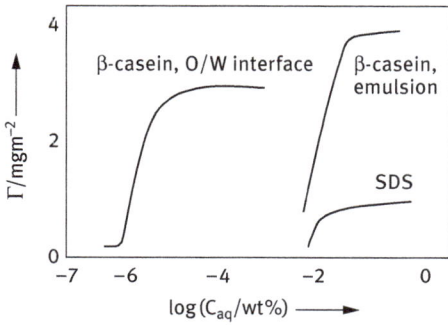

Fig. 5.15. Variation of Γ (mg m^{-2}) with log C_{eq}/wt %. The oils are β-casein (O/W interface) toluene, β-casein (emulsions) soybean, SDS benzene.

During emulsification an increase in the interfacial area A takes place and this causes a reduction in Γ. The equilibrium is restored by adsorption of surfactant from the bulk, but this takes time (shorter times occur at higher surfactant activity). Thus ϵ is small at small a and also at large a. Because of the lack or slowness of equilibrium with polymeric surfactants, ϵ will not be the same for expansion and compression of the interface.

In practice, surfactant mixtures are used and these have pronounced effects on γ and ϵ. Some specific surfactant mixtures give lower γ values than either of the two individual components. The presence of more than one surfactant molecule at the interface tends to increase ϵ at high surfactant concentrations. The various components vary in surface activity. Those with the lowest γ tend to predominate at the interface, but if present at low concentrations, it may take a long time before reaching the lowest value. Polymer-surfactant mixtures may show some synergetic surface activity.

5.3.3 Role of surfactants in droplet deformation

Apart from their effect on reducing γ, surfactants play major roles in deformation and break-up of droplets. Surfactants allow the existence of interfacial tension gradients which is crucial for formation of stable droplets. Interfacial tension gradients are very important in stabilizing the thin liquid film between the droplets which is very important during the beginning of emulsification (films of the continuous phase may be drawn through the disperse phase and collision is very large). The magnitude of the γ-gradients and of the Marangoni effect depends on the interfacial dilational modulus ϵ.

For conditions that prevail during emulsification, ϵ increases with increasing surfactant adsorption Γ and it is given by the relationship,

$$\epsilon = \frac{d\pi}{d \ln \Gamma} \tag{5.24}$$

where π is the surface pressure ($\pi = \gamma_o - \gamma$). Fig. 5.16 shows the variation of π with ln Γ; ϵ is given by the slope of the line. The SDS shows a much higher ϵ value when

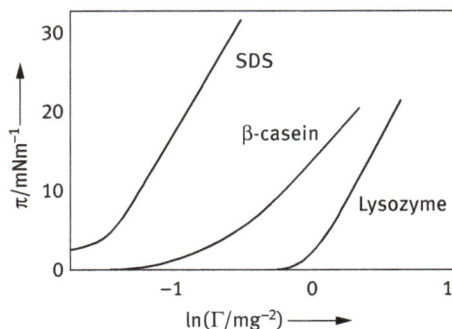

Fig. 5.16. π versus ln Γ for various emulsifiers.

compared with β-casein and lysozome. This is because the value of Γ is higher for SDS. The two proteins show a difference in their ε values which may be attributed to the conformational change that occur upon adsorption.

The presence of a surfactant means that during emulsification the interfacial tension need not be the same everywhere. This has two consequences:

1. the equilibrium shape of the drop is affected;
2. any γ-gradient formed will slow down the motion of the liquid inside the drop (this diminishes the amount of energy needed to deform and break-up the drop).

Another important role of the emulsifier is to prevent coalescence during emulsification. This is certainly not due to the strong repulsion between the droplets, since the pressure at which two drops are pressed together is much greater than the repulsive stresses. The counteracting stress must be due to the formation of γ-gradients. When two drops are pushed together, liquid will flow out from the thin layer between them, and the flow will induce a γ-gradient. This produces a counteracting stress given by

$$\tau_{\Delta\gamma} \approx \frac{2\,|\Delta\gamma|}{(1/2)d} \,.\tag{5.25}$$

The factor 2 follows from the fact that two interfaces are involved. Taking a value of $\Delta\gamma = 10$ mNm^{-1}, the stress amounts to 40 KPa (which is of the same order of magnitude as the external stress).

Closely related to the above mechanism, is the Gibbs–Marangoni effect [13–17], schematically represented in Fig. 5.17. The depletion of surfactant in the thin film between approaching drops results in γ-gradient without liquid flow being involved. This results in an inward flow of liquid that tends to drive the drops apart.

The Gibbs–Marangoni effect also explains the Bancroft rule which states that the phase in which the surfactant is most soluble forms the continuous phase. If the surfactant is in the droplets, a γ-gradient cannot develop and the drops would be prone to coalescence. Thus, surfactants with HLB > 7 tend to form O/W emulsions and HLB < 7 tend to form W/O emulsions.

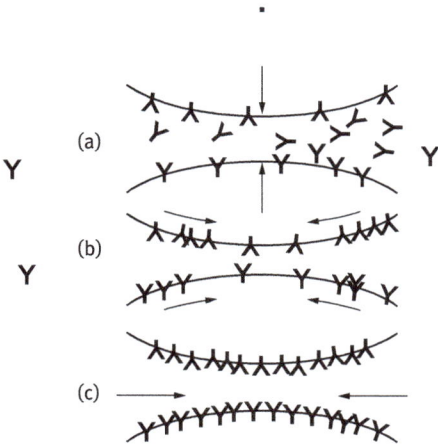

Fig. 5.17. Schematic representation of the Gibbs–Marangoni effect for two approaching drops.

The Gibbs–Marangoni effect also explains the difference between surfactants and polymers for emulsification. Polymers give larger drops when compared with surfactants. Polymers give a smaller value of ε at small concentrations when compared to surfactants.

Various other factors should also be considered for emulsification: the disperse phase volume fraction ϕ. An increase in ϕ leads to an increase in droplet collision and hence coalescence during emulsification. With increasing ϕ, the viscosity of the emulsion increases and could change the flow from being turbulent to being laminar (LV regime).

The presence of many particles results in a local increase in velocity gradients. This means that G increases. In turbulent flow, an increase in ϕ will induce turbulence depression. This will result in larger droplets. Turbulence depression by added polymers tend to remove the small eddies, resulting in the formation of larger droplets.

If the mass ratio of surfactant to continuous phase is kept constant, an increase in ϕ results in decreasing surfactant concentration and hence an increase in γ_{eq} resulting in larger droplets. If the mass ratio of surfactant to disperse phase is kept constant, the above changes are reversed.

General conclusions cannot be drawn since several of the above mentioned mechanism may come into play. Experiments using a high pressure homogenizer at various ϕ values at constant initial m_C (regime TI changing to TV at higher ϕ) showed that with increasing ϕ (> 0.1) the resulting droplet diameter increased and the dependence on energy consumption became weaker. Fig. 5.18 shows a comparison of the average droplet diameter versus power consumption using different emulsifying machines. It can be seen that the smallest droplet diameters were obtained when using the high pressure homogenizers.

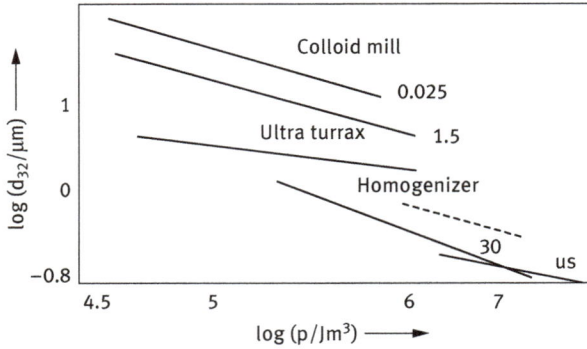

Fig. 5.18. Average droplet diameters obtained in various emulsifying machines as a function of energy consumption p. The number near the curves denote the viscosity ratio λ; the results for the homogenizer are for $\phi = 0.04$ (solid line) and $\phi = 0.3$ (dashed line); us means ultrasonic generator.

5.4 Selection of emulsifiers

5.4.1 The Hydrophilic-Lipophile Balance (HLB) concept

The selection of different surfactants in the preparation of either O/W or W/O emulsions is often still made on an empirical basis. A semi-empirical scale for selecting surfactants is the Hydrophilic-Lipophilic Balance (HLB) number developed by Griffin [18] This scale is based on the relative percentage of hydrophilic to lipophilic (hydrophobic) groups in the surfactant molecule(s). For an O/W emulsion droplet, the hydrophobic chain resides in the oil phase whereas the hydrophilic head group resides in the aqueous phase. For a W/O emulsion droplet, the hydrophilic group(s) reside in the water droplet, whereas the lipophilic groups reside in the hydrocarbon phase.

Table 5.2 gives a guide to the selection of surfactants for a particular application. The HLB number depends on the nature of the oil. As an illustration, Table 5.3 gives the required HLB numbers to emulsify various oils.

Table 5.2. Summary of HLB ranges and their applications.

HLB Range	Application
3–6	W/O emulsifier
7–9	Wetting agent
8–18	O/W emulsifier
13–15	Detergent
15–18	Solubilizer

Table 5.3. Required HLB numbers to emulsify various oils.

Oil	W/O Emulsion	O/W Emulsion
Paraffin oil	4	10
Beeswax	5	9
Lanolin, anhydrous	8	12
Cyclohexane	–	15
Toluene	–	15

The relative importance of the hydrophilic and lipophilic groups was first recognized when using mixtures of surfactants containing varying proportions of a low and high HLB number.

The efficiency of any combination (as judged by phase separation) was found to pass a maximum when the blend contained a particular proportion of the surfactant with the higher HLB number. This is illustrated in Fig. 5.19 which shows the variation of emulsion stability, droplet size and interfacial tension with % surfactant with high HLB number.

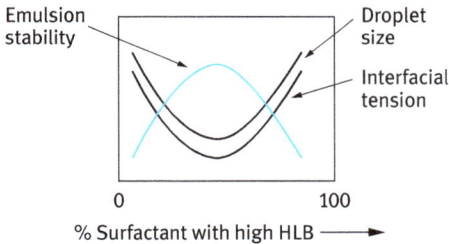

Fig. 5.19. Variation of emulsion stability, droplet size and interfacial tension with % surfactant with high HLB number.

The average HLB number may be calculated from additivity,

$$HLB = x_1 HLB_1 + x_2 HLB_2 , \tag{5.26}$$

x_1 and x_2 are the weight fractions of the two surfactants with HLB_1 and HLB_2.

Griffin developed simple equations for calculation of the HLB number of relatively simple nonionic surfactants. For a polyhydroxy fatty acid ester,

$$HLB = 20 \left(1 - \frac{S}{A} \right) , \tag{5.27}$$

S is the saponification number of the ester and A is the acid number. For a glyceryl monostearate, S = 161 and A = 198; the HLB is 3.8 (suitable for w/o emulsion).

For a simple alcohol ethoxylate, the HLB number can be calculated from the weight percent of ethylene oxide (E) and polyhydric alcohol (P),

$$HLB = \frac{E + P}{5} . \tag{5.28}$$

If the surfactant contains PEO as the only hydrophilic group, the contribution from one OH group can be neglected,

$$HLB = \frac{E}{5}.$$ (5.29)

For a nonionic surfactant $C_{12}H_{25}-O-(CH_2-CH_2-O)_6$, the HLB is 12 (suitable for O/W emulsion).

The above simple equations cannot be used for surfactants containing propylene oxide or butylene oxide. They also cannot be applied for ionic surfactants. Davies [19, 20] devised a method for calculating the HLB number for surfactants from their chemical formulae, using empirically determined group numbers. A group number is assigned to various component groups. A summary of the group numbers for some surfactants is given in Table 5.4.

Table 5.4. HLB group numbers.

Hydrophilic	Group Number
$-SO_4Na^+$	38.7
$-COO^-$	21.2
$-COONa$	19.1
N(tertiary amine)	9.4
Ester (sorbitan ring)	6.8
$-O-$	1.3
CH-(sorbitan ring)	0.5
Lipophilic	
$(-CH-), (-CH_2-), CH_3$	0.475
Derived	
$-CH_2-CH_2-O$	0.33
$-CH_2-CH_2-CH_2-O-$	−0.15

The HLB is given by the following empirical equation:

$$HLB = 7 + \sum(\text{hydrophilic group Nos}) - \sum(\text{lipohilic group Nos})\,(5).$$ (5.30)

Davies has shown that the agreement between HLB numbers calculated from the above equation and those determined experimentally is quite satisfactory.

Various other procedures were developed to obtain a rough estimate of the HLB number. Griffin found good correlation between the cloud point of 5 % solution of various ethoxylated surfactants and their HLB number.

Davies [17, 18] attempted to relate the HLB values to the selective coalescence rates of emulsions. Such correlations were not realized since it was found that the emulsion stability and even its type depend to a large extent on the method of dispersing the oil into the water and vice versa. At best the HLB number can only be used as a guide for selecting optimum compositions of emulsifying agents.

One may take any pair of emulsifying agents which fall at opposite ends of the HLB scale, e.g. Tween 80 (sorbitan monooleate with 20 moles EO, HLB = 15) and Span 80 (sorbitan monooleate, HLB = 5) using them in various proportions to cover a wide range of HLB numbers. The emulsions should be prepared in the same way, with a few percent of the emulsifying blend. The stability of the emulsions is then assessed at each HLB number from the rate of coalescence or qualitatively by measuring the rate of oil separation. In this way one may be able to find the optimum HLB number for a given oil. Having found the most effective HLB value, various other surfactant pairs are compared at this HLB value, to find the most effective pair.

5.4.2 The Phase Inversion Temperature (PIT) concept

Shinoda and coworkers [21, 22] found that many o/w emulsions stabilized with non-ionic surfactants undergo a process of inversion at a critical temperature (PIT). The PIT can be determined by following the emulsion conductivity (small amount of electrolyte is added to increase the sensitivity) as function of temperature. The conductivity of the o/w emulsion increases with increasing temperature until the PIT is reached, above which there will be a rapid reduction in conductivity (w/o emulsion is formed). Shinoda and coworkers found that the PIT is influenced by the HLB number of the surfactant. The size of the emulsion droplets was found to depend on the temperature and HLB number of the emulsifiers. The droplets are less stable towards coalescence close to the PIT. However, by rapid cooling of the emulsion a stable system may be produced. Relatively stable o/w emulsions were obtained when the PIT of the system was 20–65°C higher than the storage temperature. Emulsions prepared at a temperature just below the PIT followed by rapid cooling generally have smaller droplet sizes. This can be understood if one considers the change of interfacial tension with temperature as illustrated in Fig. 5.20. The interfacial tension decreases with increasing temperature reaching a minimum close to the PIT, after which it increases.

Thus, the droplets prepared close to the PIT are smaller than those prepared at lower temperatures. These droplets are relatively unstable towards coalescence near the PIT, but by rapid cooling of the emulsion one can retain the smaller size. This procedure may be applied to prepare mini(nano)emulsions.

The optimum stability of the emulsion was found to be relatively insensitive to changes in the HLB value or the PIT of the emulsifier, but instability was very sensitive to the PIT of the system.

It is essential, therefore, to measure the PIT of the emulsion as a whole (with all other ingredients).

At a given HLB value, stability of the emulsions against coalescence increases markedly as the molar mass of both the hydrophilic and lipophilic components in-

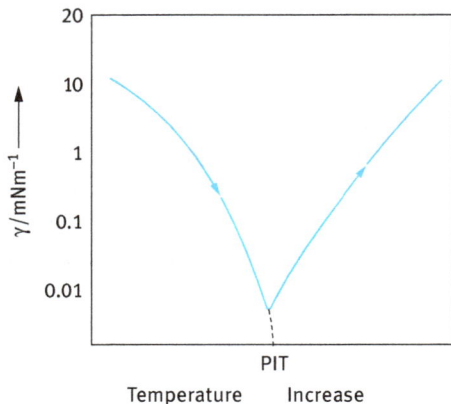

Fig. 5.20. Variation of interfacial tension with temperature increase for an O/W emulsion.

creases. The enhanced stability using high molecular weight surfactants (polymeric surfactants) can be understood from a consideration of the steric repulsion which produces more stable films. Films produced using macromolecular surfactants resist thinning and disruption thus reducing the possibility of coalescence. The emulsions showed maximum stability when the distribution of the PEO chains was broad. The cloud point is lower but the PIT is higher than in the corresponding case for narrow size distributions. The PIT and HLB number are directly related parameters.

Addition of electrolytes reduces the PIT and hence an emulsifier with a higher PIT value is required when preparing emulsions in the presence of electrolytes. Electrolytes cause dehydration of the PEO chains and in effect this reduces the cloud point of the nonionic surfactant. One needs to compensate for this effect by using a surfactant with higher HLB. The optimum PIT of the emulsifier is fixed if the storage temperature is fixed.

In view of the above correlation between PIT and HLB and the possible dependence of the kinetics of droplet coalescence on the HLB number, Sherman and coworkers suggested the use of PIT measurements as a rapid method for assessing emulsion stability. However, one should be careful in using such methods for assessment of the long-term stability since the correlations were based on a very limited number of surfactants and oils. Measurement of the PIT can at best be used as a guide for preparation of stable emulsions. Assessment of the stability should be evaluated by following the droplet size distribution as a function of time using a Coulter counter or light diffraction techniques. Following the rheology of the emulsion as a function of time and temperature may also be used for assessment of the stability against coalescence. Care should be taken in analyzing the rheological results. Coalescence results in an increase in the droplet size and this is usually followed by a reduction in the viscosity of the emulsion. This trend is only observed if the coalescence is not accompanied by flocculation of the emulsion droplets.

5.5 Stabilization of emulsions

5.5.1 Creaming or sedimentation and its prevention

This is the result of gravity, when the density of the droplets and the medium are not equal. Fig. 5.1 gives a schematic picture for creaming or sedimentation. For small droplets ($< 0.1\,\mu$, i.e. nanoemulsions) where the Brownian diffusion kT (where k is the Boltzmann constant and T is the absolute temperature) exceeds the force of gravity (mass x acceleration due to gravity g),

$$kT \ll \frac{4}{3}\pi R^3 \Delta\rho\, g\, L, \qquad (5.31)$$

where R is the droplet radius, $\Delta\rho$ is the density difference between the droplets and the medium and L is the height of the container, no creaming or sedimentation occurs. However, with most emulsions, with a size distribution in the range 0.1–5 μm (with an average of ~1–2 μm), the gravity force is much higher than the Brownian diffusion and in this case the droplets will cream or sediment at various rates. In the last case, a concentration gradient builds up with the larger droplets staying at the top of the cream layer or the bottom,

$$C(h) = C_o \exp\left(-\frac{mgh}{kT}\right) \qquad (5.32)$$

$$m = \frac{4}{3}\pi R^3 \Delta\rho\, g, \qquad (5.33)$$

C(h) is the concentration (or volume fraction ϕ) of droplets at height h, whereas C_o is the concentration at the top or bottom of the container.

For very dilute emulsions ($\phi \leq 0.01$) the rate could be calculated using Stokes' law which balances the hydrodynamic force with gravity force,

$$v_o \frac{2}{9}\frac{\Delta\rho\, g\, R^2}{\eta_o}, \qquad (5.34)$$

v_o is the Stokes velocity and η_o is the viscosity of the medium.

For an O/W emulsion with $\Delta\rho = 0.2$ in water ($\eta_o \sim 10^{-3}$ Pas), the rate of creaming or sedimentation is $\sim 4.4 \times 10^{-5}$ ms^{-1} for 10 μm droplets and $\sim 4.4 \times 0^{-7}$ ms^{-1} for 1 μm droplets. This means that in a 0.1 m container creaming or sedimentation of the 10 μm droplets is complete in ~ 0.6 hour and for the 1 μm droplets this takes ~ 60 hours.

For moderately concentrated emulsions ($0.2 > \phi > 0.1$), one has to take into account the hydrodynamic interaction between the droplets, which reduces the Stokes velocity to a value v given by the following expression,

$$v = v_o(1 - k\phi), \qquad (5.35)$$

where k is a constant that accounts for hydrodynamic interaction. k is of the order of 6.5, which means that the rate of creaming or sedimentation is reduced by about 65 %.

For more concentrated emulsions (ϕ > 0), the rate of creaming or sedimentation becomes a complex function of ϕ. v decreases with increasing ϕ and ultimately it approaches zero when ϕ exceeds a critical value, ϕ_p, which is the so-called "maximum packing fraction". The value of ϕ_p for monodisperse "hard-spheres" ranges from 0.64 (for random packing) to 0.74 for hexagonal packing. The value of ϕ_p exceeds 0.74 for polydisperse systems. Also for emulsions which are deformable, ϕ_p can be much larger than 0.74. When ϕ approaches ϕ_p, the relative viscosity η_r approaches ∞. In practice most emulsions are prepared at ϕ values well below ϕ_p, usually in the range 0.2–0.5, and under these conditions creaming or sedimentation is the rule rather than the exception.

Several procedures may be applied to reduce or eliminate creaming or sedimentation and these are discussed below.

(1) Matching density of oil and aqueous phases:
Clearly if $\Delta\rho$ = 0, v = 0. However, this method is seldom practical. Density matching, if possible, only occurs at one temperature.

(2) Reduction of droplet size:
Since the gravity force is proportional to R^3, then if R is reduced by a factor of 10, the gravity force is reduced by 1000. Below a certain droplet size (which also depends on the density difference between oil and water), the Brownian diffusion may exceed gravity and creaming or sedimentation is prevented. This is the principle of formulation of nanoemulsions (with size range 50–200 nm) which may show very little or no creaming or sedimentation. The same applies for microemulsions (size range 5–50 nm)

(3) Use of "thickeners":
These are high molecular weight polymers, natural or synthetic such as Xanthan gum, hydroxyethyl cellulose, alginates, carrageenans, etc. These "thickeners" have very high viscosities at low stresses or shear rates (to be denoted as the residual or zero shear viscosity $\eta(o)$ above a critical polymer concentration (C^*) which can be located from plots of $\log \eta$ versus $\log C$). In most cases good correlation between the rate of creaming or sedimentation and $\eta(o)$ is obtained.

5.5.2 Flocculation of emulsions and its prevention

As mentioned above, flocculation is the result of van der Waals attraction that is universal for all disperse systems. As mentioned before, to overcome the van der Waals attraction one can use electrostatic stabilization using ionic surfactants which results in the formation of electrical double layers that introduce a repulsive energy that overcomes the attractive energy. Emulsions stabilized by electrostatic repulsion become flocculated at intermediate electrolyte concentrations. The second and most effective method of overcoming flocculation is by "steric stabilization" using nonionic surfactants or polymers. Stability may be maintained in electrolyte solutions (as high as 1 mol dm^{-3} depending on the nature of the electrolyte) and up to high temperatures (in excess of 50°C) provided the stabilizing chains (e.g. PEO) are still in better than θ-conditions ($\chi < 0.5$).

5.5.3 Ostwald ripening and its reduction

The driving force for Ostwald ripening is the difference in solubility between the small and large droplets (the smaller droplets have higher Laplace pressure and higher solubility than the larger ones). The difference in chemical potential between different sized droplets was given by Lord Kelvin [23],

$$S(r) = S(\infty) \exp\left(\frac{2\gamma V_m}{r\,RT}\right), \tag{5.36}$$

where $S(r)$ is the solubility surrounding a particle of radius r, $S(\infty)$ is the bulk solubility, V_m is the molar volume of the dispersed phase, R is the gas constant and T is the absolute temperature. The quantity $(2\gamma V_m/rRT)$ is termed the characteristic length. It has an order of ~1nm or less, indicating that the difference in solubility of a 1 μm droplet is of the order of 0.1 % or less. Theoretically, Ostwald ripening should lead to condensation of all droplets into a single drop. This does not occur in practice since the rate of growth decreases with increasing droplet size.

For two droplets with radii r_1 and r_2 ($r_1 < r_2$),

$$\frac{RT}{V_m} \ln\left[\frac{S(r_1)}{S(r_2)}\right] = 2\gamma\left[\frac{1}{r_1} - \frac{1}{r_2}\right]. \tag{5.37}$$

Equation (5.37) shows that the larger the difference between r_1 and r_2, the higher the rate of Ostwald ripening. Ostwald ripening can be quantitatively assessed from plots of the cube of the radius versus time t [24–26],

$$r^3 = \frac{8}{9}\left[\frac{S(\infty)\,\gamma\,V_m\,D}{\rho\,RT}\right]t, \tag{5.38}$$

D is the diffusion coefficient of the disperse phase in the continuous phase.

Several methods may be applied to reduce Ostwald ripening:

1. Addition of a second disperse phase component which is insoluble in the continuous medium (e.g. squalane) [27]. In this case partitioning between different droplet sizes occurs, with the component having low solubility expected to be concentrated in the smaller droplets. During Ostwald ripening in a two component system, equilibrium is established when the difference in chemical potential between different size droplets (which results from curvature effects) is balanced by the difference in chemical potential resulting from partitioning of the two components. This effect reduces further growth of droplets.

2. Modification of the interfacial film at the O/W interface. According to equation
    (5.38) reduction in γ results in a reduction of the Ostwald ripening rate. By using
    surfactants that are strongly adsorbed at the O/W interface (i.e. polymeric surfac-
    tants) and which do not desorb during ripening (by choosing a molecule that is
    insoluble in the continuous phase) the rate could be significantly reduced [28]. An
    increase in the surface dilational modulus ϵ (= dy/dln A) and decrease in γ would
    be observed for the shrinking drop and this tends to reduce further growth.

A–B–A block copolymers such as PHS–PEO–PHS (which is soluble in the oil droplets
but insoluble in water) can be used to achieve the above effect. This polymeric emulsifier enhances the Gibbs elasticity and causes reduction of γ to very low values.

5.5.4 Emulsion coalescence and its prevention

When two emulsion droplets come in close contact in a floc or creamed layer or during
Brownian diffusion, thinning and disruption of the liquid film may occur resulting in
eventual rupture. On close approach of the droplets, film thickness fluctuations may
occur. Alternatively, the liquid surfaces undergo some fluctuations forming surface
waves. The surface waves may grow in amplitude and the apices may join as a result
of the strong van der Waals attraction (at the apex, the film thickness is the smallest).
The same applies if the film thins to a small value (critical thickness for coalescence)
    A very useful concept was introduced by Deryaguin [29] who suggested that a "Dis-
joining Pressure" $\pi(h)$ is produced in the film which balances the excess normal pressure,

$$\pi(h) = P(h) - P_o,\tag{5.39}$$

where $P(h)$ is the pressure of a film with thickness h and P_o is the pressure of a sufficiently thick film such that the net interaction free energy is zero.
    $\pi(h)$ may be equated to the net force (or energy) per unit area acting across the
film,

$$\pi(h) = -\frac{dG_T}{dh},\tag{5.40}$$

where G_T is the total interaction energy in the film.
    $\pi(h)$ is made of three contributions due to electrostatic repulsion (π_E), steric re-
pulsion (π_S) and van der Waals attraction (π_A),

$$\pi(h) = \pi_E + \pi_S + \pi_A.\tag{5.41}$$

To produce a stable film $\pi_E + \pi_S > \pi_A$ and this is the driving force for prevention of
coalescence which can be achieved by two mechanisms and their combination:

1. Increased repulsion both electrostatic and steric.
2. Dampening of the fluctuation by enhancing the Gibbs elasticity. In general, smaller droplets are less susceptible to surface fluctuations and hence coalescence is reduced. This explains the high stability of nanoemulsions.

Several methods may be applied to achieve the above effects:

(1) Use of mixed surfactant films:
In many cases using mixed surfactants, say anionic and nonionic or long chain alcohols, can reduce coalescence as a result of several effects: high Gibbs elasticity; high surface viscosity; hindered diffusion of surfactant molecules from the film.

(2) Formation of lamellar liquid crystalline phases at the O/W interface:
This mechanism was suggested by Friberg and coworkers [30], who suggested that surfactant or mixed surfactant film can produce several bilayers that "wrap" the droplets. As a result of these multilayer structures, the potential drop is shifted to longer distances thus reducing the van der Waals attraction. For coalescence to occur, these multilayers have to be removed "two-by-two" and this forms an energy barrier preventing coalescence.

References

[1] Th. F. Tadros and B. Vincent, in: *Encyclopedia of Emulsion Technology*, P. Becher (ed.), Marcel Dekker, New York, 1983.
[2] B. P. Binks (ed.), *Modern Aspects of Emulsion Science*, The Royal Society of Chemistry Publication, Cambridge, 1998.
[3] Th. Tadros, *Applied Surfactants*, Wiley-VCH, Germany 2005.
[4] H. C. Hamaker, *Physica* (Utrecht) **4**, 1058 (1937).
[5] B. V. Deryaguin and L. Landua, *Acta Physicochem.*, USSR **14**, 633 (1941).
[6] E. J. W. Verwey and J. Th. G. Overbeek, *Theory of Stability of Lyophobic Colloids*, Elsevier, Amsterdam, 1948.
[7] D. H. Napper, *Polymeric Stabilisation of Dispersions*, Academic Press, London, 1983.
[8] P. Walstra and P. E. A. Smolders, in: *Modern Aspects of Emulsions*, B. P. Binks (ed.), The Royal Society of Chemistry, Cambridge, 1998.
[9] H. A. Stone, *Ann. Rev. Fluid Mech.*, **226**, 95 (1994).
[10] J. A. Wierenga, F. ven Dieren, J. J. M. Janssen and W. G. M. Agterof, *Trans. Inst. Chem. Eng.*, **74-A**, 554 (1996).
[11] V. G. Levich, *Physicochemical Hydrodynamics*, Prentice-Hall, Englewood Cliffs, 1962.
[12] J. T. Davis, *Turbulent Phenomena*, Academic Press, London, 1972.
[13] E. H. Lucasses-Reynders, in: *Encyclopedia of Emulsion Technology*, P. Becher (ed.), Marcel Dekker, New York, 1996.
[14] D. E. Graham and M.C. Phillips, *J. Colloid Interface Sci.*, **70**, 415 (1979).
[15] E. H. Lucasses-Reynders, *Colloids and Surfaces*, **A91**, 79 (1994).
[16] J. Lucassen, in: *Anionic Surfactants*, E.H. Lucassesn-Reynders (ed.), Marcel Dekker, New York, 1981.

[17]   M. van den Tempel, *Proc. Int. Congr. Surf. Act.*, **2**, 573 (1960).

[18]   W. C. Griffin, *J. Cosmet. Chemists*, **1**, 311 (1949); **5**, 249 (1954).

[19]   J. T. Davies, *Proc. Int. Congr. Surface Activity*, Vol. 1, p 426 (1959).

[20]   J. T. Davies and E. K. Rideal, *Interfacial Phenomena*, Academic Press, New York, 1961.

[21]   K. Shinoda, *J. Colloid Interface Sci.*, **25**, 396 (1967).

[22]   K. Shinoda and H. Saito, *J. Colloid Interface Sci.*, **30**, 258 (1969).

[23]   W. Thompson (Lord Kelvin), *Phil. Mag.*, **42**, 448 (1871).

[24]   A. S. Kabalanov and E. D. Shchukin, *Adv. Colloid Interface Sci.*, **38**, 69 (1992). A. S. Kabalanov, *Langmuir*, **10**, 680 (1994).

[25]   I. M. Lifshitz and V. V. Slesov, *Sov. Phys. JETP*, **35**, 331 (1959).

[26]   C. Wagner, *Z. Electrochem.*, **35**, 581 (1961).

[27]   W. I. Higuchi and J. Misra, *J. Pharm. Sci.*, **51**, 459 (1962).

[28]   P. Walstra, in: *Encyclopedia of Emulsion Technology*, P. Becher (ed.), Marcel Dekker, New York, 1996.

[29]   B. V. Deryaguin and R. L. Scherbaker, *Kolloid Zh.*, **23**, 33 (1961).

[30]   S. Friberg, P. O. Jansson and E. Cederberg, *J. Colloid Interface Sci.*, **55**, 614 (1976).

6 Surfactants as dispersants and stabilization of suspensions

6.1 Introduction

For dispersion of powders in liquids and stabilization of suspensions it is necessary to add surfactants. The same applies for the preparation of suspensions by condensation methods starting from molecular units. For these reasons, surfactants find application in almost every industrial preparation, e.g., paints, dyestuffs, paper coatings, printing inks, agrochemicals, pharmaceuticals, cosmetics, food products, detergents, ceramics, etc. For preparation of suspensions from preformed materials that are supplied as powders, surfactants are essential ingredients and the final product is determined by the nature and amount of the surfactant added. The powder can be hydrophobic, e.g. organic pigments, agrochemicals, ceramics or hydrophilic, e.g. silica, titania, clays. The liquid can be aqueous or non-aqueous. The role of surfactants in dispersing solids in liquids can be understood from their accumulation at the solid/liquid interface. This was described in detail in Chapter 4. It is essential to understand the process of dispersion at a fundamental level: "Dispersion is a process whereby aggregates and agglomerates of powders are dispersed into 'individual' units, usually followed by a wet milling process (to subdivide the particles into smaller units) and stabilization of the resulting dispersion against aggregation and sedimentation" [1, 2].

In this chapter, I will describe the role of surfactants in the preparation of solid/liquid dispersions (suspensions). The stabilization of suspensions by surfactants both electrostatically and setrically will be briefly described.

6.2 Role of surfactants in preparation of solid/liquid dispersions (suspensions)

There are two main processes for the preparation of solid/liquid dispersions. The first depends on the "build-up" of particles from molecular units, i.e. the so-called condensation method, which involves two main processes, namely nucleation and growth. In this case, it is necessary first to prepare a molecular (ionic, atomic or molecular) distribution of the insoluble substances; then by changing the conditions precipitation is caused leading to the formation of nuclei that grow to the particles in question. In the second procedure, usually referred to as a dispersion process, larger "lumps" of the insoluble substances are subdivided by mechanical or other means into smaller units. The role of surfactants in the preparation of suspensions by these two methods will be described separately.

6.2.1 Role of surfactants in condensation methods. Nucleation and growth

To understand the role of surfactants in the condensation methods, it is essential to consider the major processes involved, namely nucleation and growth. Nucleation is the spontaneous process of the appearance of a new phase from a metastable (supersaturated) solution of the material in question [3]. The initial stages of nucleation result in the formation of small nuclei where the surface to volume ratio is very large and hence the role of specific surface energy is very important. With the progressive increase of the size of the nuclei, the ratio becomes smaller and eventually large crystals appear, with a corresponding reduction in the role played by the specific surface energy. As we will see later, addition of surfactants can be used to control the process of nucleation and the size of the resulting nucleus.

According to Gibbs [4] and Volmer [5], the free energy of formation of a spherical nucleus, ΔG, is given by the sum of two contributions: a positive surface energy term ΔG_s which increases with an increase in the radius of the nucleus r, and a negative contribution ΔG_v due to the appearance of a new phase, which also increases with increasing r,

$$\Delta G = \Delta G_s + \Delta G_v , \tag{6.1}$$

ΔG_s is given by the product of area of the nucleus and the specific surface energy (solid/liquid interfacial tension) σ; ΔG_v is related to the relative supersaturation (S/S_o),

$$\Delta G = 4\pi r^2 \gamma - \left(\frac{4\pi r^3 \rho}{3M} \right) RT \ln \left(\frac{S}{S_o} \right) , \tag{6.2}$$

ρ is the density, R is the gas constant and T is the absolute temperature.

In the initial stages of nucleation, ΔG_s increases faster with increasing r when compared to ΔG_v and ΔG remains positive, reaching a maximum at a critical radius r^*, after which it decreases and eventually becomes negative. This occurs since the second term in equation (6.2) rises faster with increasing r than the first term (r^3 versus r^2). When ΔG becomes negative, growth becomes spontaneous and the cluster grows very fast. This is illustrated in Fig. 6.1. This figure shows the critical size of the nucleus r^* above which growth becomes spontaneous. The free energy maximum ΔG^* at the critical radius represents the barrier that has to be overcome before growth becomes spontaneous. Both r^* and ΔG^* can be obtained by differentiating equation (6.2) with respect to r and equating the result to zero. This gives the following expressions:

$$r^* = \frac{2\gamma M}{\rho RT \ln(S/S_o)} \tag{6.3}$$

$$\Delta G^* = \frac{16}{3} \frac{\pi \gamma^3 M^2}{(\rho RT)^2 [(\ln(S/S_o)]^2} . \tag{6.4}$$

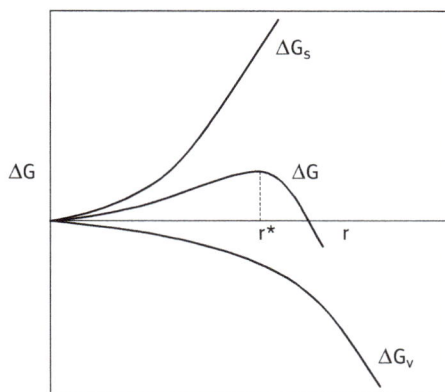

Fig. 6.1. Variation of free energy of formation of a nucleus with radius r.

It is clear from equations (6.1)–(6.4) that the free energy of formation of a nucleus and the critical radius r^* above which the cluster formation grows spontaneously depend on two main parameters, σ and (S/S_0), both of which are influenced by the presence of surfactants. σ is influenced in a direct way by adsorption of surfactant on the surface of the nucleus; this adsorption lowers σ and this reduces r^* and ΔG^*. In other words, spontaneous formation of clusters occurs at smaller critical radius. In addition, surfactant adsorption stabilizes the nuclei against any flocculation. The presence of micelles in solution also affects the process of nucleation and growth directly and indirectly. The micelles can act as "nuclei" on which growth may occur. In addition, the micelles may solubilize the molecules of the material, thus affecting the relative supersaturation and this can have an effect on nucleation and growth.

6.2.2 Emulsion polymerization

In emulsion polymerization, the monomer, e.g. styrene or methyl methacrylate that is insoluble in the continuous phase, is emulsified using a surfactant that adsorbs at the monomer/water interface [6]. The surfactant micelles in bulk solution solubilize some of the monomer. A water-soluble initiator such as potassium persulfate $K_2S_2O_8$ is added and this decomposes in the aqueous phase forming free radicals that interact with the monomers forming oligomeric chains. It has long been assumed that nucleation occurs in the "monomer swollen micelles". The reasoning behind this mechanism was the sharp increase in the rate of reaction above the critical micelle concentration and that the number of particles formed and their size depend to a large extent on the nature of the surfactant and its concentration (which determines the number of micelles formed). However, later this mechanism was disputed and it was suggested that the presence of micelles means that excess surfactant is available and molecules will readily diffuse to any interface.

The most accepted theory of emulsion polymerization is referred to as the coagulative nucleation theory [7, 8]. A two-step coagulative nucleation model has been proposed by Napper and coworkers [7, 8]. In this process the oligomers grow by propagation and this is followed by a termination process in the continuous phase. A random coil is produced which is insoluble in the medium and this produces a precursor oligomer at the θ-point. The precursor particles subsequently grow primarily by coagulation to form true latex particles. Some growth may also occur by further polymerization. The colloidal instability of the precursor particles may arise from their small size, and the slow rate of polymerization can be due to reduced swelling of the particles by the hydrophilic monomer [7, 8]. The role of surfactants in these processes is crucial since they determine the stabilizing efficiency and the effectiveness of the surface active agent ultimately determines the number of particles formed. This was confirmed by using surface active agents of different nature. The effectiveness of any surface active agent in stabilizing the particles was the dominant factor and the number of micelles formed was relatively unimportant.

According to the theory of Smith and Ewart [9] of the kinetics of emulsion polymerization, the rate of propagation R_p is related to the number of particles N formed in a reaction by the equation,

$$-\frac{d[M]}{dt} = R_p \, k_p \, N \, n_{av}[M] \tag{6.5}$$

where [M] is the monomer concentration in the particles, k_p is the propagation rate constant and n_{av} is the average number of radicals per particle.

According to equation (6.5), the rate of polymerization and the number of particles are directly related to each other, i.e. an increase in the number of particles will increase the rate. This has been found for many polymerizations, although here are some exceptions. The number of particles is related to the surfactant concentration [S] by the equation [8],

$$N approx [S]^{3/5} \tag{6.6}$$

Using the coagulative nucleation model, Napper et al. [7, 8] found that the final particle number increases with increasing surfactant concentration with a monotonically diminishing exponent. The slope of $d(\log N_c)/d(\log t)$ varies from 0.4 to 1.2. At high surfactant concentration, the nucleation time will be long in duration since the new precursor particles will be readily stabilized. As a result, more latex particles are formed and eventually will outnumber the very small precursor particles at long times. The precursor/particle collisions will become more frequent and fewer latex particles are produced. The dN_c/dt will approach zero and at long times the number of latex particles remains constant. This shows the inadequacy of the Smith–Ewart theory which predicts a constant exponent (3/5) at all surfactant concentrations. For this reason, the coagulative nucleation mechanism has now been accepted as the most probable theory for emulsion polymerization. In all cases, the nature and concentration of surfactant used is very crucial and this is very important in the industrial preparation of latex systems.

Most reports on emulsion polymerization have been limited to commercially available surfactants, which in many cases are relatively simple molecules such as sodium dodecyl sulfate and simple nonionic surfactants. However, studies on the effect of surfactant structure on latex formation have revealed the importance of the structure of the molecule. Block and graft copolymers (polymeric surfactants) are expected to be better stabilizers when compared to simple surfactants. Studies on styrene polymerization using an A-B block of polystyrene with polyethylene oxide (PS-PEO) with various ratios of the molecular weight of the two blocks showed that an optimum composition is required [10]. For efficient anchoring to the latex particles, the block length need not be more than 10 units and the PEO block with a molecular weight of 3000 was sufficient to stabilize the particles. The results also showed that using a higher molecular weight stabilizer could be counterproductive.

6.2.3 Dispersion polymerization

In this case the reaction mixture consisting of monomer, initiator and solvent (aqueous or nonaqueous) for both are usually homogeneous, but as polymerization proceeds, polymer separates out and the reaction continues in a homogeneous manner [11]. A dispersant, sometimes referred to as "protective agent", is added to stabilize the particles once formed.

The above mechanism for the preparation of polymer particles is usually applied for preparation of nonaqueous dispersions (latex particles dispersed in a nonaqueous medium), referred to as nonaqueous dispersion polymerization (NAD). As mentioned above, the two main criteria for this type of polymerization are the insolubility of the formed polymer in the continuous phase and the solubility of the monomer and initiator in the dispersion medium. Initially the polymerization starts as a homogeneous system, but after polymerization proceeds to some extent, the insolubility of the formed polymer chains causes their precipitation. The process can be visualized as starting with the formation of polymer chains by free radical initiation, followed by formation of nuclei which then grow into polymer particles.

In the early production of nonaqueous latex dispersions, the continuous medium was chosen to be a hydrocarbon solvent. However, later mixed solvents with polar components were used. Indeed, the process of dispersion polymerization has been applied in many cases using completely polar solvents such as alcohol, or alcohol-water mixtures [11].

The mechanism of dispersion polymerization has been discussed in detail in the book edited by Barrett [11]. A distinct difference between emulsion and dispersion polymerization may be considered in terms of the rate of reaction. As mentioned above, with emulsion polymerization the rate of reaction depends on the number of particles formed. However, with dispersion polymerization, the rate is independent of the number of particles formed. This is to be expected, since in the latter case polymerization initially occurs in the continuous phase where both monomer and initiator are soluble, and the continuation of polymerization after precipitation is questionable. Although in emulsion polymerization the initial monomer initiation reaction also occurs in the continuous medium, the particles formed become swollen with the monomer and polymerization may continue in these particles. A comparison of the rate of reaction for dispersion and solution polymerization showed a much faster rate for the former process [11].

As mentioned above, to prevent aggregation of the formed polymer particles one needs a dispersant (polymer surfactant) which must satisfy a number of criteria. The most effective dispersants are those of the block (A–B or A–B–A) or graft (BA$_n$) type. The B chain is chosen to be insoluble in the medium and has high affinity to the surface of the polymer particles (or become incorporated within its matrix). This is usually referred to as the "anchor" chain. The A chain(s) are chosen to be highly soluble in the medium and strongly solvated with its molecules. The solvation of the chain and its solubility (in a good solvent for A) is described by the Flory–Huggins interaction parameter χ, which in a good solvent is <0.5, in order to ensure effective steric stabilization.

The nature and concentration of the stabilizer determines the number of particles formed in dispersion polymerization. In general increasing dispersant concentration increases the number of particles formed (at any given monomer content), i.e. smaller latex particles are produced. This is not surprising since smaller particles have larger surface area and this requires a higher dispersant concentration.

The particles in dispersion polymerization were considered to be formed by two main steps [10]:
1. initiation of monomer in the continuous phase and subsequent growth of the oligomeric chains until insolubility occurs;
2. growing oligomeric chains associate with each other to form aggregates, which below a certain critical size are unstable but gain stability through dispersant adsorption.

However, several other processes may take place, e.g. homocoagulation (collision with other precursor particles), growth by propagation, adsorption of stabilizer and swelling by monomer. It should be pointed out, however, that the number of particles in the final latex cannot be dependent on particle nucleation only, since there is another step involved which determines how many of the precursor particles created are involved in the formation of one colloidally stable particle. This step depends on the nature of the stabilizer and how many particles have to heterocoagulate to decrease the total surface area to a size that the stabilizer in the system is capable of stabilizing.

6.2.4 Role of surfactants in dispersion methods

As mentioned before, dispersion methods are used for the preparation of suspension of preformed particles. The term dispersion is used to refer to the complete process of incorporating the solid into a liquid such that the final product consists of fine particles distributed throughout the dispersion medium. The role of surfactants (or polymeric surfactants) in the dispersion can be seen from consideration of the stages involved [1]. Three stages have been considered [3]: wetting of the powder by the liquid, breaking of the aggregates and agglomerates and comminution (milling) of the resulting particles into smaller units.

Wetting is a fundamental process in which one fluid phase is displaced completely or partially by another fluid phase from the surface of a solid. A useful parameter to describe wetting is the contact angle θ of a liquid drop on a solid substrate. If the liquid makes no contact with the solid, i.e. $\theta = 180°$, the solid is referred to as non-wettable by the liquid in question. This may be the case for a perfectly hydrophobic surface with a polar liquid such as water. However, when $180° > \theta > 90°$, one may refer to a case of poor wetting. When $0° < \theta < 90°$, partial (incomplete) wetting is the case, whereas when $\theta = 0°$ complete wetting occurs and the liquid spreads on the solid substrate forming a uniform liquid film. The utility of contact angle measurements depends on equilibrium thermodynamic arguments (static measurements) using the well-known Young's equation [12]. The value depends on:
1. the history of the system;
2. whether the liquid is tending to advance across or recede from the solid surface (Advancing angle θ_A, Receding angle θ_R; usually $\theta_A > \theta_R$).

Under equilibrium, the liquid drop takes the shape that minimizes the free energy of the system. Three interfacial tensions can be identified: γ_{SV}, Solid/Vapor area A_{SV} – γ_{SL}, Solid/Liquid area A_{SL} – γ_{LV}, Liquid/Vapor area A_{LV}. A schematic representation of the balance of tensions at the solid/liquid/vapor interface is shown in Fig. 6.2. The contact angle is that formed between the planes tangent to the surfaces of the solid and liquid at the wetting perimeter. Here, solid and liquid are simultaneously in contact with each other and the surrounding phase (air or vapor of the liquid). The wetting perimeter is referred to as the three phase line or wetting line. In this region there is an equilibrium between vapor, liquid and solid.

$\gamma_{SV}A_{SV} + \gamma_{SL}A_{SL} + \gamma_{LV}A_{LV}$ should be a minimum at equilibrium and this leads to the well-known Young's equation,

$$\gamma_{SV} = \gamma_{SL} + \gamma_{LV} \cos \theta \tag{6.7}$$

$$\cos \theta = \frac{\gamma_{SV} - \gamma_{SL}}{\gamma_{LV}} . \tag{6.8}$$

The contact angle θ depends on the balance between the solid/vapor (γ_{SV}) and solid/liquid (γ_{SL}) interfacial tensions. The angle which a drop assumes on a solid

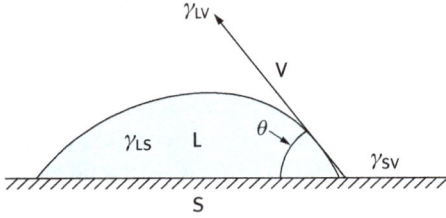

Fig. 6.2. Schematic representation of the contact angle.

surface is the result of the balance between the adhesion force between solid and liquid and the cohesive force in the liquid,

$$\gamma_{LV} \cos \theta = \gamma_{SV} - \gamma_{SL} . \tag{6.9}$$

If there is no interaction between solid and liquid,

$$\gamma_{SL} = \gamma_{SV} + \gamma_{LV} , \tag{6.10}$$

i.e., $\cos \theta = -1$ or $\theta = 180°$.

If there is strong interaction between solid and liquid (maximum wetting), the latter spreads until Young's equation is satisfied,

$$\gamma_{LV} = \gamma_{SV} - \gamma_{SL} , \tag{6.11}$$

i.e., $\cos \theta = 1$ or $\theta = 0°$; the liquid is described to spread spontaneously on the solid surface.

When the surface of the solid is in equilibrium with the liquid vapor, we can consider the spreading pressure π; the solid surface tension is lowered as a result of adsorption of vapor molecules

$$\pi = \gamma_s - \gamma_{SV} . \tag{6.12}$$

Young's equation can be written as:

$$\gamma_{LV} \cos \theta = \gamma_s - \gamma_{SL} - \pi . \tag{6.13}$$

There is no direct way by which γ_{SV} or γ_{SL} can be measured. The difference between γ_{SV} and γ_{SL} can be obtained from contact angle measurements ($= \gamma_{LV} \cos \theta$). This difference is referred to as: Wetting Tension or Adhesion Tension

$$\text{Adhesion Tension} = \gamma_{SV} - \gamma_{SL} = \gamma_{LV} \cos \theta . \tag{6.14}$$

Gibbs defined the adhesion tension τ as the difference between the surface pressure of the solid/liquid and that between the solid/vapor interface,

$$\tau = \pi_{SL} - \pi_{SV} \tag{6.15}$$

$$\pi_{SV} = \gamma_s - \gamma_{SV} \tag{6.16}$$

$$\pi_{SL} = \gamma_s - \gamma_{SL} \tag{6.17}$$

$$\tau = \gamma_{SV} - \gamma_{SL} = \gamma_{LV} \cos \theta . \tag{6.18}$$

The work of adhesion is a direct measure of the free energy of interaction between solid and liquid,

$$W_a = (\gamma_{LV} + \gamma_{SV}) - \gamma_{SL}. \tag{6.19}$$

Using Young's equation,

$$W_a = \gamma_{LV} + \gamma_{SV} - \gamma_{LV} \cos\theta = \gamma_{LV}(\cos\theta + 1). \tag{6.20}$$

The work of adhesion depends on: γ_{LV}, the liquid/vapor surface tension and θ, the contact angle between liquid and solid.

The work of cohesion W_c is the work of adhesion when the two surfaces are the same,

$$W_c = 2\gamma_{LV}. \tag{6.21}$$

For adhesion of a liquid on a solid, $W_a \sim W_c$ or $\theta = 0°$ ($\cos\theta = 1$).

Harkins [12] defined the spreading coefficient as the work required to destroy unit area of SL and LV and leaves unit area of bare solid SV, i.e.,

Spreading coefficient S = Surface energy of final state – Surface energy of initial state

$$S = \gamma_{SV} - (\gamma_{SL} + \gamma_{LV}). \tag{6.22}$$

Using Young's equation,

$$\gamma_{SV} = \gamma_{SL} + \gamma_{LV} \cos\theta \tag{6.23}$$

$$S = \gamma_{LV}(\cos\theta - 1). \tag{6.24}$$

If S is zero (or positive), i.e. $\theta = 0$, the liquid will spread until it completely wets the solid. If S is negative, i.e. $\theta > 0$, only partial wetting occurs. Alternatively, one can use the equilibrium (final) spreading coefficient.

For dispersion of powders into liquids, one usually requires complete spreading, i.e. θ should be zero.

For a liquid spreading on a uniform, non-deformable solid (idealized case), there is only one contact angle-equilibrium value. With real systems (practical solids) a number of stable contact angles can be measured. Two relatively reproducible angles can be measured: largest-advancing angle θ_A; smallest-receding angle θ_R. θ_A is measured by advancing the periphery of a drop over a surface (e.g. by adding more liquid to the drop). θ_R is measured by pulling the liquid back. $(\theta_A - \theta_R)$ is referred to as contact angle hysteresis. The latter is caused by three main factors:
1. Penetration of wetting liquid into pores during advancing contact angle measurements.
2. Surface roughness: the first and rear edges both meet the liquid with some intrinsic angle θ_0 (microscopic contact angle). The macroscopic angles θ_A and θ_R vary significantly. This is best illustrated for a surface inclined at an angle α from the horizontal. θ_0 values are determined by contact of liquid with the "rough" valleys

(microscopic contact angle). θ_A and θ_R are determined by contact of liquid with arbitrary parts on the surface (peak or valley). Surface roughness can be accounted for by comparing the "real" area of the surface A with that of the projected,

$$r = \frac{A}{A'}, \tag{6.25}$$

A = area of surface taking into account all peaks and valleys. A' = Apparent area (same macroscopic dimension); $r > 1$. θ is related to θ_0 by the Wenzel equation,

$$\cos\theta = r\cos\theta_0, \tag{6.26}$$

θ = Macroscopic contact angle,
θ_0 = Microscopic contact angle

$$\cos\theta = r\left[\frac{(\gamma_{SV} - \gamma_{SL})}{\gamma_{LV}}\right]. \tag{6.27}$$

If $\cos\theta$ is negative on a smooth surface ($\theta > 90°$), it becomes more negative on a rough surface (θ is larger) and surface roughness reduces wetting. If $\cos\theta$ is positive on a smooth surface ($\theta < 90°$), it becomes more positive on a rough surface (θ is smaller) and roughness enhances wetting.

3. Surface heterogeneity: most practical surfaces are heterogeneous consisting of "islands" or "patches" with different surface energies. As the drop advances on such a surface, the edge of the drop tends to stop at the boundary of the "island". The advancing angle will be associated with the intrinsic angle of the high contact angle region. The receding angle will be associated with the low contact angle region. If the heterogeneities are very small compared with the dimensions of the liquid drop, one can define a composite contact angle using Cassie's equation,

$$\cos\theta = Q_1 \cos\theta_1 + Q_2 \cos\theta_2, \tag{6.28}$$

Q_1 = Fraction of surface having contact angle θ_1; Q_2 = Fraction of surface having contact angle θ_2. θ_1 and θ_2 are the maximum and minimum possible angles.

Surfactants lower the surface tension of water, γ, and they adsorb at the solid/liquid interface. As mentioned in Chapter 4, the plot of γ_{LV} versus log C (where C is the surfactant concentration) results in a gradual reduction in γ_{LV} followed by a linear decrease of γ_{LV} with log C (just below the critical micelle concentration, cmc) and when the cmc is reached γ_{LV} remains virtually constant. From the slope of the linear portion of the γ–log C curve (just below the cmc), one can obtain the surface excess (number of moles of surfactant per unit area at the L/A interface). Using the Gibbs adsorption isotherm,

$$\frac{d\gamma}{d\log C} = -2.303\, RT\,\Gamma, \tag{6.29}$$

Γ = surface excess (moles m^{-2}); R = gas constant; T = absolute temperature.

From Γ one can obtain the area per molecule,

$$\text{Area per molecule} = \frac{1}{\Gamma\,N_{av}}\ (m^2) = \frac{10^{18}}{\Gamma\,N_{av}}\ (nm^2). \tag{6.30}$$

Most surfactants produce a vertically-oriented monolayer just below the cmc. The area/molecule is usually determined by the cross-sectional area of the head group. For ionic surfactants containing say $-OSO_3^-$ or $-SO_3^-$ head group, the area per molecule is in the region of $0.4\ nm^2$. For nonionic surfactants containing several moles of ethylene oxide (8-10), the area per molecule can be much larger ($1-2\ nm^2$). Surfactants will also adsorb at the solid/liquid interface. For hydrophobic surfaces, the main driving force for adsorption is by hydrophobic bonding. This results in lowering of the contact angle of water on the solid surface. For hydrophilic surfaces, adsorption occurs via the hydrophilic group, e.g. cationic surfactants on silica. Initially the surface becomes more hydrophobic and the contact angle θ increases with increase in surfactant concentration. However, at higher cationic surfactant concentration, a bilayer is formed by hydrophobic interaction between the alkyl groups and the surface becomes more and more hydrophilic and eventually the contact angle reaches zero at high surfactant concentrations.

Smolders [14] suggested the following relationship for change of θ with C,

$$\frac{d\gamma_{LV}\cos\theta}{d\ln C} = \frac{d\gamma_{SV}}{d\ln C} - \frac{d\gamma_{SL}}{d\ln C}. \tag{6.31}$$

Using the Gibbs equation,

$$\sin\theta\left(\frac{d\gamma}{d\ln C}\right) = RT\,(\Gamma_{SV} - \Gamma_{SL} - \gamma_{LV}\cos\theta) \tag{6.32}$$

since $\gamma_{LV}\sin\theta$ is always positive, then $(d\theta/d\ln C)$ will always have the same sign as the RHS of equation (6.32). Three cases may be distinguished: $(d\theta/d\ln C) < 0$; $\Gamma_{SV} < \Gamma_{SL} + \Gamma_{LV}\cos\theta$; addition of surfactant improves wetting. $(d\theta/d\ln C) = 0$; $\Gamma_{SV} = \Gamma_{SL} + \Gamma_{LV}\cos\theta$; surfactant has no effect on wetting. $(d\theta/d\ln C) > 0$; $\Gamma_{SV} > \Gamma_{SL} + \Gamma_{LV}\cos\theta$; surfactant causes dewetting.

Wetting of powders by liquids is very important in their dispersion, e.g. in the preparation of concentrated suspensions. The particles in a dry powder form either aggregates (where the particles are connected by their surfaces) or agglomerates (where the particles are connected by their corners). It is essential in the dispersion process to wet both external and internal surfaces and displace the air entrapped between the particles. Wetting is achieved by the use of surface active agents (wetting agents) of the ionic or nonionic type which are capable of diffusing quickly (i.e. lower the dynamic surface tension) to the solid/liquid interface and displace the air entrapped by rapid penetration through the channels between the particles and inside any "capillaries". For wetting of hydrophobic powders into water, anionic surfactants, e.g. alkyl sulfates or sulfonates or nonionic surfactants of the alcohol or alkyl phenol ethoxylates are usually used.

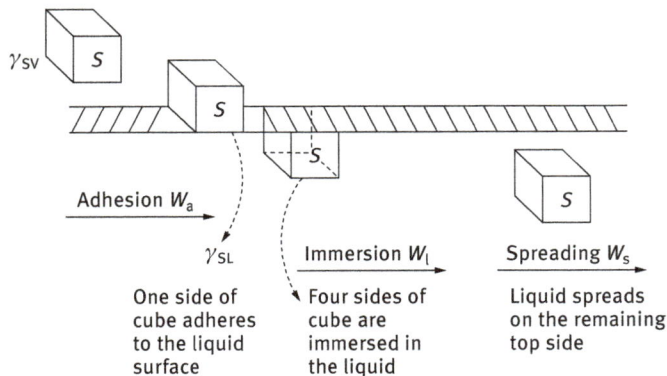

Fig. 6.3. Schematic representation of wetting of a cube of solid.

The process of wetting of a solid by a liquid involves three types of wetting: Adhesion wetting, W_a; Immersion wetting W_i; Spreading wetting W_s. This can be illustrated by considering a cube of solid with unit area of each side (Fig. 6.3). In every step one can apply the Young's equation,

$$\gamma_{SV} = \gamma_{SL} + \gamma_{LV} \cos \theta \tag{6.33}$$

$$W_a = \gamma_{SL} - (\gamma_{SV} + \gamma_{LV}) = -\gamma_{LV}(\cos \theta + 1) \tag{6.34}$$

$$W_i = 4\gamma_{SL} - 4\gamma_{SV} = -4\gamma_{LV} \cos \theta \tag{6.35}$$

$$W_s = (\gamma_{SL} + \gamma_{LV}) - \gamma_{SV} = -\gamma_{LV}(\cos \theta - 1) . \tag{6.36}$$

The work of dispersion W_d is the sum of W_a, W_i and W_s,

$$W_d = W_a + W_i + W_s = 6\gamma_{SV} - \gamma_{SL} = -6\gamma_{LV} \cos \theta . \tag{6.37}$$

Wetting and dispersion depend on: γ_{LV}, liquid surface tension; θ, contact angle between liquid and solid. W_a, W_i and W_s are spontaneous when $\theta < 90°$. W_d is spontaneous when $\theta = 0$. Since surfactants are added in sufficient amounts ($\gamma_{dynamic}$ is lowered sufficiently) spontaneous dispersion is the rule rather than the exception.

Wetting of the internal surface requires penetration of the liquid into channels between and inside the agglomerates. The process is similar to forcing a liquid through fine capillaries. To force a liquid through a capillary with radius r, a pressure p is required that is given by

$$p = -\frac{2\gamma_{LV} \cos \theta}{r} = \left[\frac{-2(\gamma_{SV} - \gamma_{SL})}{r\gamma_{LV}} \right], \tag{6.38}$$

γ_{SL} has to be made as small as possible; rapid surfactant adsorption to the solid surface, low θ. When $\theta = 0$, $p \propto \gamma_{LV}$. Thus for penetration into pores one requires a high γ_{LV}. Thus, wetting of the external surface requires low contact angle θ and low surface tension γLV. Wetting of the internal surface (i.e. penetration through pores) requires low θ but high γ_{LV}. These two conditions are incompatible and a compromise has to be made: $\gamma_{SV} - \gamma_{SL}$ must be kept at a maximum. γ_{LV} should be kept as low as possible but not too low.

The above conclusions illustrate the problem of choosing the best dispersing agent for a particular powder. This requires measurement of the above parameters as well as testing the efficiency of the dispersion process.

The rate of liquid penetration is described by the Rideal–Washburn equation [15, 16]

$$l = \left[\frac{r\, t\, \gamma_{LV} \cos\theta}{2\eta} \right]^{1/2},$$
(6.39)

where l is the depth of penetration at time t, r is the capillary radius and η is the viscosity of the medium.

To enhance the rate of penetration, γ_{LV} has to be made as high as possible, θ as low as possible and η as low as possible. For dispersion of powders into liquids one should use surfactants that lower θ while not reducing γ_{LV} too much. The viscosity of the liquid should also be kept at a minimum. Thickening agents (such as polymers) should not be added during the dispersion process. It is also necessary to avoid foam formation during the dispersion process.

For a packed bed of particles, r may be replaced by k, which contains the effective radius of the bed and a turtuosity factor, which takes into account the complex path formed by the channels between the particles, i.e.,

$$l^2 = \frac{k\, t\, \gamma_{LV} \cos\theta}{2\eta}.$$
(6.40)

Thus a plot of l^2 versus t gives a straight line and from the slope of the line one can obtain θ.

The Rideal–Washburn equation can be applied to obtain the contact angle of liquids (and surfactant solutions) in powder beds. K should first be obtained using a liquid that produces zero contact angle. This is discussed below.

6.3 Assessment of wettability of powders

6.3.1 Sinking time, submersion or immersion test

This by far the most simple (but qualitative) method for assessment of wettability of a powder by a surfactant solution. The time for which a powder floats on the surface of a liquid before sinking into the liquid is measured. 100 ml of the surfactant solution is placed in a 250 ml beaker (of internal diameter of 6.5 cm) and after 30 min. standing 0.30 g of loose powder (previously screened through a 200-mesh sieve) is distributed with a spoon onto the surface of the solution. The time t for the 1 to 2 mm thin powder

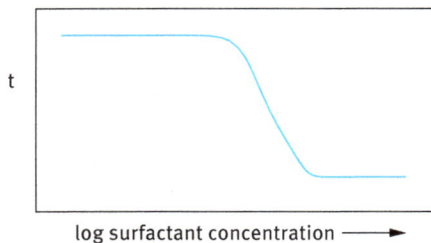

Fig. 6.4. Sinking time as a function of surfactant concentration.

layer to completely disappear from the surface is measured using a stop watch. Surfactant solutions with different concentrations are used and t is plotted versus surfactant concentration as illustrated in Fig. 6.4.

The lower the surfactant concentration at which the sinking time shows a sharp decrease, the better the wetting agent.

6.3.2 Measurement of contact angles of liquids and surfactant solutions on powders

The contact angle value gives a more quantitative measurement of wetting; the lower the value the better the wetting agent. A special procedure is used for measurement of the contact angle on powders. A packed bed of the powder is prepared, say in a tube fitted with a sintered glass at the end (to retain the powder particles). It is essential to pack the powder uniformly in the tube (a plunger may be used in this case). The tube containing the bed is immersed in a liquid that gives spontaneous wetting (e.g. a lower alkane), i.e. the liquid gives a zero contact angle and $\cos \theta = 1$. By measuring the rate of penetration of the liquid (this can be carried out gravimetrically using for example a microbalance or a Kruss instrument) one can obtain k. The tube is then removed from the lower alkane liquid and left to stand for evaporation of the liquid. It is then immersed in the liquid in question and the rate of penetration is measured again as a function of time. Using equation (6.40), one can calculate $\cos \theta$ and hence θ.

6.3.3 List of wetting agents for hydrophobic solids in water

The most effective wetting agent is the one that gives a zero contact angle at the lowest concentration. For $\theta = 0°$ or $\cos \theta = 1$, γ_{SL} and γ_{LV} have to be as low as possible. This requires quick reduction of γ_{SL} and γ_{LV} under dynamic conditions during powder dispersion (this reduction should normally be achieved in less than 20 seconds). This requires fast adsorption of the surfactant molecules both at the L/V and S/L interfaces. It should be mentioned that reduction of γ_{LV} is not always accompanied by simultaneous reduction of γ_{SL} and hence it is necessary to have information on both interfacial tensions which means that measurement of the contact angle is essential in selection of wetting agents. Measurement of γ_{SL} and γ_{LV} should be carried out under dynamic conditions (i.e. at very short times). In the absence of such measurements, the sinking time described above could be applied as a guide for wetting agent selection. The most commonly used wetting agents for hydrophobic solids are listed below.

To achieve rapid adsorption the wetting agent should be either a branched chain with central hydrophilic group or a short hydrophobic chain with hydrophilic end group. The most commonly used wetting agents are the following:

Aerosol OT (diethylhexyl sulfosuccinate)

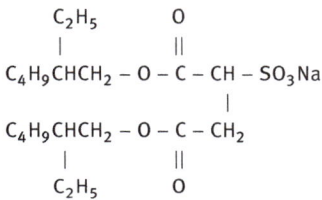

$$
\begin{array}{cc}
C_2H_5 & O \\
| & || \\
C_4H_9CHCH_2 - O - C - CH - SO_3Na \\
& | \\
C_4H_9CHCH_2 - O - C - CH_2 \\
| & || \\
C_2H_5 & O
\end{array}
$$

The above molecule has a low critical micelle concentration (cmc) of 0.7 gdm-3 and at and above the cmc the water surface tension is reduced to ~ 25 mNm-1 in less than 15 s.

An alternative anionic wetting agent is sodium dodecylbenzene sulfonate with a branched alkyl chain

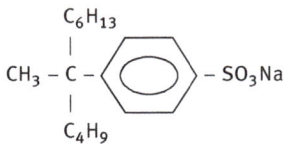

$$
\begin{array}{c}
C_6H_{13} \\
| \\
CH_3 - C - \bigcirc\!\!\!\!\bigcirc - SO_3Na \\
| \\
C_4H_9
\end{array}
$$

The above molecule has a higher cmc (1 gdm-3) than Aerosol OT. It is also not as effective in lowering the surface tension of water reaching a value of 30 mNm-1 at and above the cmc. It is, therefore, not as effective as Aerosol OT for powder wetting.

Several nonionic surfactants such as the alcohol ethoxylates can also be used as wetting agents. These molecules consist of a short hydrophobic chain (mostly C10) which is also branched. A medium chain polyethylene oxide (PEO) mostly consisting of 6 EO units or lower is used. These molecules also reduce the dynamic surface tension within a short time (< 20 s) and they have reasonably low cmc. In all cases one should use the minimum amount of wetting agent to avoid interference with the dispersant that needs to be added to maintain the colloid stability during dispersion and on storage.

6.3.4 Stabilization of suspensions using surfactants

The stabilization of suspensions with surfactants is determined by the balance between the van der Waals attraction and electrostatic and/or steric repulsion between the particles containing adsorbed surfactant molecules. These interaction forces have been described in Chapter 5 and one can schematically describe the three energy-distance curves (Figs. 6.5–6.7) for electrostatic, steric and electrosteric (combination of electrostatic and repulsive forces). Fig. 6.5 shows the case for electrostatically stabilized dispersions according to the Deryaguin–Landau–Verwey–Overbeek (DLVO) theory [17, 18] at low electrolyte concentration ($< 10^{-2}$ mol dm^{-3} 1:1 electrolyte, e.g. NaCl). In this case the total interaction G_T is the sum of electrostatic repulsion G_e and van der Waals attraction G_A. At long distances of separation, $G_A > G_e$, resulting in a shallow minimum (secondary minimum). At very short distances, $G_A \gg G_e$, resulting in a deep primary minimum. At intermediate distances, $G_e > G_A$, resulting in energy maximum, G_{max}, whose height depends on the surface potential ψ_0 (or Stern potential ψ_d) and the electrolyte concentration and valency. Fig. 6.6 shows the case of sterically stabilized dispersions (when using nonionic surfactants) [19] where the interaction is the sum of G_{mix} (unfavorable mixing of the stabilizing chains when in good solvent conditions), G_{el} (entropic or elastic interaction resulting from reduction of configurational entropy of the chains on considerable overlap) and G_A. G_{mix} increases very sharply with decreasing h, when h < 2δ. G_{el} increases very sharply with decreasing h,

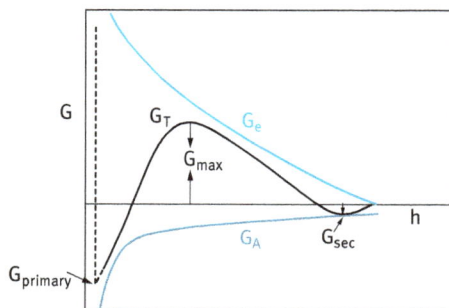

Fig. 6.5. Energy-distance curves for electrostatically stabilized dispersions.

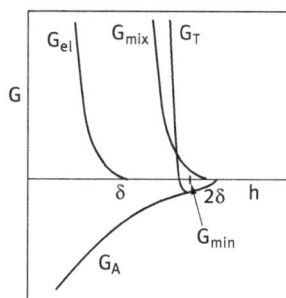

Fig. 6.6. Energy-distance curves for sterically stabilized systems.

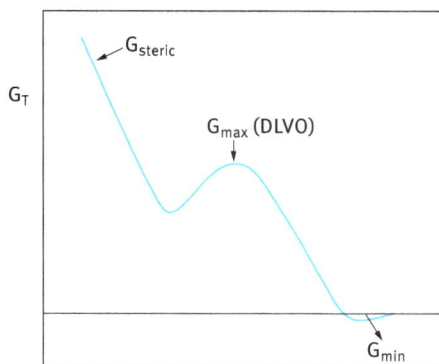

Fig. 6.7. Energy-distance curves for electrosterically stabilized systems.

when $h < \delta$. G_T versus h shows a minimum, G_{min}, at separation distances comparable to 2δ. When $h < 2\delta$, G_T shows a rapid increase with decreasing h. The depth of the minimum depends on the Hamaker constant A, the particle radius R and adsorbed layer thickness δ. G_{min} increases with increasing A and R. At a given A and R, G_{min} increases with decreasing δ (i.e. with a decrease of the molecular weight, M_w, of the stabilizer). Fig. 6.7 shows the case of electrosteric repulsion (for example when using a mixture of ionic and nonionic surfactant) where G_T is the sum of G_e, G_s $(G_{mix} + G_{el})$ and G_A. In this case the energy-distance curve has two minima, one shallow maximum (corresponding to the DLVO type) and a rapid increase at small h corresponding to steric repulsion.

References

[1]   Th. Tadros, *Dispersions of Powders in Liquids and Stabilization of Suspensions*, Wiley-VCH, Germany, 2012.

[2]   Th. F. Tadros (ed.), *Solid/Liquid Dispersions*, Academic Press, London, 1987.

[3]   G. D. Parfitt (ed.), *Fundamental Aspects of Dispersions, in Dispersion of Powders in Liquids*, Applied Science Publishers, London, 1973.

[4]   J. W. Gibbs, *Scientific Papers*, Vol. 1, Longman Green, London, 1906,

[5]   M. Volmer, *Kinetik der Phase Buildung*, Stemkopf, Dresden, 1939.

[6]   D. Blakely, *Emulsion Polymerization*, Applied Science Publication, London, 1975.

[7]   G. Litchi, R. G. Gilbert and D. H. Napper, *J. Polym. Sci.*, **21**, 269 (1983).

[8]   P. J. Feeney, D. H. Napper and R. G. Gilbert, *Macromolecules*, **17**, 2520 (1984).

[9]   W. V. Smith and R. H. Ewart, *J. Chem. Phys.*, **16**, 592 (1948).

[10]   I. Piirma, *Polymeric Surfactants*, Surfactant Science Series, Vol 42, Marcel Dekker, New York, 1992.

[11]   K. E. J. Barrett, *Dispersion Polymerization in Non-Aqueous Media*, John Wiley and Sons, London, 1975.

[12]   T. Blake, Wetting, in: *Surfactants*, Th. F. Tadros (ed.), Academic Press, London, 1984.

[13]   W. A. Zisman, *Contact Angles, Wettability and Adhesion*, Advances in Chemistry Series, No. 43, p. 1, ACS, Wasington, 1964.

[14]   C. A. Smolders, *Rec. Trav. Chim.* **80** 650 (1961).

[15]   E. K. Rideal, *Phil. Mag.*, **44**, 1152 (1922).

[16]   E. D. Washburn, *Phys. Rev.*, **17**, 273 (1921).

[17]   B. V. Deryaguin and L. Landau, *Acta Physicochim. USSR,* **14**, 633 (1941).

[18]   E. J. W. Verwey and J. Th. G. Overbeek, *Theory of Stability of Lyophobic Colloids*, Elsevier, Amsterdam, 1948.

[19]   D. H. Napper, *Polymeric Stabilisation of Colloidal Dispersions*, Academic Press, London 1983.

7 Surfactants for foam stabilization

7.1 Introduction

Foam is a disperse system, consisting of gas bubbles separated by liquid layers. It can be simply produced when air or some other gas is introduced beneath the surface of a liquid that expands to enclose the gas with a film of liquid. Because of the significant density difference between the gas bubbles and the medium, the system quickly separates into two layers with the gas bubbles rising to the top, which may undergo deformation to form polyhedral structures. Pure liquids cannot foam unless a surface active material is present. When a gas bubble is introduced below the surface of a pure liquid, it burst almost immediately as a soon as the liquid has drained away. With dilute surfactant solutions, as the liquid/air interface expands and the equilibrium at the surface is disturbed, a resorting force is set up which tries to establish the equilibrium. The restoring force arises from the Gibb–Marangoni effect which was discussed in detail in Chapter 5. As a result of the presence of surface tension gradients dγ (due to incomplete coverage of the film by surfactant), a dilational elasticity ε is produced (Gibbs elasticity). This surface tension gradient induces flow of surfactant molecules from the bulk to the interface and these molecules carry liquid with them (the Marangoni effect). The Gibbs–Marangoni effect prevent thinning and disruption of the liquid film between the air bubbles and this stabilizes the foam. This process will be discussed in detail below.

Several surface active foaming materials may be distinguished:
1. Surfactants: ionic, nonionic and zwitterionic.
2. Polymers (polymeric surfactants).
3. Particles that accumulate at the air/solution interface.
4. Specifically adsorbed cations or anions from inorganic salts. Many of these substances can cause foaming at extremely low concentrations (as low as 10^{-9} mol dm^{-3}).

In kinetic terms foams may be classified into:
1. Unstable, transient foams (lifetime of seconds).
2. Metastable, permanent foams (lifetimes of hours or days).

7.2 Foam preparation

Like most disperse systems, foams can be obtained by condensation and dispersion methods. The condensation method for generating foam involves creation of gas bubbles in the solution by decreasing the external pressure, by increasing temperature or as a result of chemical reaction. Thus, bubble formation may occur through homogeneous nucleation that occurs at high supersaturation or heterogeneous nucleation (e.g. from catalytic sites) that occurs at low supersaturation. The most applied technique for generating foam is by a simple dispersion technique (mechanical shaking or whipping). This method is not satisfactory, since accurate control of the amount of air incorporated is difficult to achieve. The most convenient method is to pass a flow of gas (sparging) through an orifice with well-defined radius r_o.

The size of the bubbles produced at an orifice, r, may be roughly estimated from the balance of the buoyancy force F_b with the surface tension force F_s [1],

$$F_b = (4/3)\pi r^3 \rho g \tag{7.1}$$

$$F_s = 2\pi r_o \gamma \tag{7.2}$$

$$r = \left(\frac{3\gamma r_o}{2\rho g} \right)^{1/3}, \tag{7.3}$$

r and r_o are the radii of the bubble and orifice and ρ is the specific gravity of liquid.

Since the dynamic surface tension of the growing bubble is higher than the equilibrium tension, the contact base may spread, depending on the wetting conditions. Thus, the main problem is the value of γ to be used in equation (7.3). Another important factor that controls bubble size is the adhesion tension $\gamma \cos\theta$, where θ is the dynamic contact angle of the liquid on the solid of the orifice. With a hydrophobic surface, a bubble develops with a greater size than the hole. One should always distinguish between the equilibrium contact angle θ and the dynamic contact angle, θ_{dyn} during bubble growth. As the bubble detaches from the orifice, the dimensions of the bubble will determine the velocity of the rise. The rise of the bubble through the liquid causes a redistribution of surfactant on the bubble surface, with the top having a reduced concentration and the polar base having a higher concentration than the equilibrium value. This unequal distribution of surfactant on the bubble surface has an important role in foam stabilization (due to the surface tension gradients). When the bubble reaches the interface, a thin liquid film is produced on its top. The lifetime of this thin film depends on many factors, e.g. surfactant concentration, rate of drainage, surface tension gradient, surface diffusion and external disturbances.

7.3 Foam structure

Two main types of foams may be distinguished:
1. Spherical foam ("Kugel Schaum") consisting of gas bubbles separated by thick films of viscous liquid produced in freshly prepared systems. This may be considered as a temporary dilute dispersion of bubbles in the liquid.
2. Polyhedral gas cells produced on aging; thin flat "walls" are produced with junction points of the interconnecting channels (Plateau borders). Due to the interfacial curvature, the pressure is lower and the film is thicker in the plateau border.

A capillary suction effect of the liquid occurs from the center of the film to its periphery.

The pressure difference between neighboring cells, Δp, is related to the radius of curvature (r) of the Plateau border by

$$\Delta p = \frac{2\gamma}{r} \,.$$ (7.4)

In a foam column, several transitional structures may be distinguished. Near the surface, a high gas content (polyhedral foam) is formed, whereas near the base of the column a much lower gas content structure is obtained that is referred to as the bubble zone. A transition state may be distinguished between the upper and bottom layers. The drainage of excess liquid from the foam column to the underlying solution is initially driven by hydrostatic, which causes the bubble to become distorted. The foam collapse usually occurs from top to bottom of the column. The films in the polyhedral foam are more susceptible to rupture by shock, temperature gradient or vibration. Another mechanism of foam instability is due to Ostwald ripening (disproportionation). The driving force for this process is the difference in Laplace pressure between the small and the larger foam bubble. The smaller bubbles have higher Laplace pressure than the larger ones. The gas solubility increases with pressure and hence gas molecules will diffuse from the smaller to the larger bubbles. This process only occurs with spherical foam bubbles. This process may be opposed by the Gibbs elasticity effect. Alternatively, rigid films produced using polymers may resist Ostwald ripening as a result of the high surface viscosity. With polyhedral foam with planer liquid lamella, the pressure difference between the bubbles is not large and hence Ostwald ripening is not the mechanism for foam instability in this case. With polyhedral foam, the main driving force for foam collapse is the surface forces that act across the liquid lamella. To keep the foam stable (i.e. to prevent complete rupture of the film), this capillary suction effect must be prevented by an opposing "disjoining pressure" that acts between the parallel layers of the central flat film (see below). The generalized model for drainage involves the Plateau borders forming a "network" through which the liquid flows due to gravity.

7.4 Classification of foam stability

All foams are thermodynamically unstable (due to the high interfacial free energy). As mentioned above, foams are classified according to the kinetics of their breakdown:
1. Unstable (transient) foams, lifetime seconds. These are generally produced using "mild" surfactants, e.g. short chain alcohols, aniline, phenol, pine oil, short chain undissociated fatty acid. Most of these compounds are sparingly soluble and may produce a low degree of elasticity.

2. Metastable ("permanent") foams, lifetime hours or days. These metastable foams are capable of withstanding ordinary disturbances (thermal or Brownian fluctuations). They can collapse from abnormal disturbances (evaporation, temperature gradients, etc.)

The above metastable foams are produced from surfactant solutions near or above the critical micelle concentration (cmc). The stability is governed by the balance of surface forces (see below). The film thickness is comparable to the range of intermolecular forces. In the absence of external disturbances, these foams may stay stable indefinitely. They are produced using proteins, long chain fatty acids or solid particles.

Gravity is the main driving force for foam collapse, directly or indirectly through the Plateau border. Thinning and disruption may be opposed by surface tension gradients at the air/water interface. Alternatively the drainage rate may be decreased by increasing the bulk viscosity of the liquid (e.g. addition of glycerol or polymers). Stability may be increased in some cases by the addition of electrolytes that produce a "gel network" in the surfactant film. Foam stability may also be enhanced by increasing the surface viscosity and/or surface elasticity. High packing of surfactant films (high cohesive forces) may also be produced using mixed surfactant films or surfactant/polymer mixtures.

For investigation of foam stability one must consider the role of the Plateau border under dynamic and static conditions. One should also consider foam films with intermediate lifetimes, i.e. between unstable and metastable foams.

7.4.1 Drainage and thinning of foam films

As mentioned above, gravity is the main driving force for film drainage. Gravity can act directly on the film or through capillary suction in the Plateau borders. As a general rule, the rate of drainage of foam films may be decreased by increasing the bulk viscosity of the liquid from which the foam is prepared. This can be achieved by adding glycerol or high molecular weight polymer such as poly(ethylene oxide). Alternatively, the viscosity of the aqueous surfactant phase can be increased by addition of electrolytes that form a "gel" network (liquid crystalline phases may be produced). Film drainage can also be decreased by increasing the surface viscosity and surface elasticity. This can be achieved, for example, by addition of proteins, polysaccharides and even particles. These systems are applied in many food foams.

Most quantitative studies on film drainage have been carried out using small, horizontal films as described in detail by Scheludko and coworkers [2–4]. The film thickness is determined by interferometry, which is based on comparison between the intensities of the light falling on the film and that reflected from it [4]. For thinner films, large electrostatic repulsive interactions can reduce the driving force for drainage and may lead to stable films. Also, for thick films containing high surfactant concentrations (> the critical micelle concentration, cmc), the micelles present in the film can cause a repulsive structural mechanism. The effect of deformation of the film surface during thinning is also extremely complicated.

Foam films can also be produced by pulling a frame out of a reservoir containing a surfactant solution [5, 6]. Three stages can be identified:
1. Initial formation of the film that is determined by the withdraw velocity;
2. Drainage of the film within the lamella which causes thinning with time;
3. Aging of the film that may result in the formation of a metastable film.

Assuming that the monolayer of the surfactant film at the boundaries of the film is rigid, film drainage may be described by the viscous flow of the liquid under gravity between two parallel plates, as given by Poiseille's equation,

$$V_{av} = \frac{\rho\, g\, h^2}{8\eta},\tag{7.5}$$

where h is the film thickness, ρ is the liquid density in the film, η is the viscosity of the liquid and g is the acceleration due to gravity.

As the process proceeds, the thinning can also occur by a horizontal mechanism known as marginal regeneration [7, 8] in which the liquid is drained from the film near the border region and exchanged from within the low pressure plateau border. In this exchange, the total area of the film does not change significantly. This regeneration mechanism results in the formation of patches of thin film at the border, with the excess fluid flowing into the border channel. The edge effect determines the drainage, with the rate of thinning varying inversely with film width [7–9]. This results in thickness fluctuations caused by capillary waves. Marginal regeneration is probably the most important cause of drainage in vertical films with mobile surfaces, i.e. with surfactant solutions at concentrations above the cmc.

7.4.2 Theories of foam stability

There is no single theory that can explain foam stability in a satisfactory manner. Several approaches have been considered and these are summarized below.

7.4.2.1 Surface viscosity and elasticity theory
The adsorbed surfactant film is assumed to control the mechanical-dynamical properties of the surface layers by virtue of its surface viscosity and elasticity. This concept may be true for thick films (> 100 nm) where intermolecular forces are less dominant (i.e. foam stability under dynamic conditions). Surface viscosity reflects the speed of the relaxation process which restores the equilibrium in the system after imposing a

stress on it. Surface elasticity is a measure of the energy stored in the surface layer as a result of an external stress.

The viscoelastic properties of the surface layer are an important parameter. The most useful technique to study the viscoelastic properties of surfactant monolayers is the surface scattering methods. When transversal ripples occur, periodic dilation and compression of the monolayer occurs and this can be accurately measured. This enables one to obtain the viscoelastic behavior of monolayers under equilibrium and non-equilibrium conditions, without disturbing the original state of the adsorbed layer. Some correlations have been found between surface viscosity and elasticity and foam stability, e.g. when adding lauryl alcohol to sodium lauryl sulfate which tends to increase the surface viscosity and elasticity [10].

7.4.2.2 The Gibbs–Marangoni effect theory

The Gibbs coefficient of elasticity, ϵ, was introduced as a variable resistance to surface deformation during thinning:

$$\epsilon = 2 \left(\frac{d\gamma}{d \ln A} \right) = 2 \left(\frac{d\gamma}{d \ln h} \right),$$ (7.6)

$d \ln h$ = relative change in lamella thickness, ϵ is the "film elasticity of compression modulus" or "Surface Dilational Modulus". ϵ is a measure of the ability of the film to adjust its surface tension in an instant stress. In general, the higher the value of ϵ the more stable the film is. ϵ depends on surface concentration and film thickness. For a freshly produced film to survive, a minimum ϵ is required.

The main deficiency of the early studies on Gibbs elasticity was that it was applied to thin films and the diffusion from the bulk solution was neglected. In other words, the Gibbs theory applies to the case where there are insufficient surfactant molecules in the film to diffuse to the surface and lower the surface tension. This is clearly not the case with most surfactant films. For thick lamella under dynamic conditions, one should consider diffusion from the bulk solution, i.e. the Marangoni effect. The Marangoni effect tends to oppose any rapid displacement of the surface (Gibbs effect) and may provide a temporary restoring force to "dangerous" thin films. In fact, the Marangoni effect is superimposed on the Gibbs elasticity, so that the effective restoring force is a function of the rate of extension, as well as the thickness. When the surface layers behave as insoluble monolayers, then the surface elasticity has its greatest value and is referred to as the Marangoni dilational modulus, ϵ_m.

The Gibbs–Marangoni effect explains the maximum foaming behavior at intermediate surfactant concentration [5]. At low surfactant concentrations (well below the cmc), the greatest possible differential surface tension will only be relatively small and little foaming will occur. At very high surfactant concentration (well above the cmc), the differential tension relaxes too rapidly because of the supply of surfactant which diffuses to the surface. This causes the restoring force to have time to counteract the disturbing forces and produces a dangerously thinner film and foaming is poor. It is the intermediate surfactant concentration range that produces maximum foaming.

7.4.2.3 Surface forces theory (disjoining pressure π)

This theory operates under static (equilibrium) conditions in relatively dilute surfactant solutions (h <100 nm). In the early stages of formation, foam films drain under the action of gravitation or capillary forces. Provided the films remain stable during this drainage stage, they may approach a thickness in the range of 100 nm. At this stage, surface forces come into play, i.e. the range of the surface forces becomes now comparable to the film thickness. Deryaguin and coworkers [11, 12] introduced the concept of disjoining pressure which should remain positive to slow down further drainage and film collapse. This is the principle of formation of thin metastable (equilibrium) films.

In addition to the Laplace capillary pressure, three additional forces can operate at surfactant concentration below the cmc: electrostatic double layer repulsion π_{el}, van der Waals attraction π_{vdW} and steric (short range) forces π_{st},

$$\pi = \pi_{el} + \pi_{vdw} + \pi_{st} . \tag{7.7}$$

In the original definition of disjoining pressure by Deryaguin [11, 12], he only considered the first two terms on the right-hand side of equation (7.7). At low electrolyte concentrations, double layer repulsion predominates and π_{el} can compensate the capillary pressure, i.e. $\pi_{el} = P_c$. This results in the formation of an equilibrium-free film which is usually referred to as the thick common film CF (∼50 nm thickness). This equilibrium metastable film persists until thermal or mechanical fluctuations cause rupture. The stability of the CF can be described in terms of the theory of colloid stability due to Deryaguin, Landau [13] and Verwey and Overbeek [14] (DLVO theory).

The critical thickness value at which the CF ruptures (due to thickness perturbations) fluctuates and an average value h_{cr} may be defined. However, an alternative situation may occur as h_{cr} is reached and instead of rupturing, a metastable film (high stability) may be formed with a thickness $h < h_{cr}$. The formation of this metastable film can be experimentally observed through the formation of "islands of spots" which appear black in light reflected from the surface. This film is often referred to a "first black" or "common black" film. The surfactant concentration at which this "first black" film is produced can be 1–2 orders of magnitude lower than the cmc.

Further thinning can cause an additional transformation into a thinner stable region (a stepwise transformation). This usually occurs at high electrolyte concentrations which leads to a second, very stable, thin black film usually referred to as Newton secondary black film, with a thickness in the region of 4 nm. Under these conditions, the short range steric or hydration forces control the stability and this provides the third contribution to the disjoining press, π_{st} described in equation (7.7).

Fig. 7.1 shows a schematic representation of the variation of disjoining pressure π with film thickness h, which shows the transition from the common film to the com-

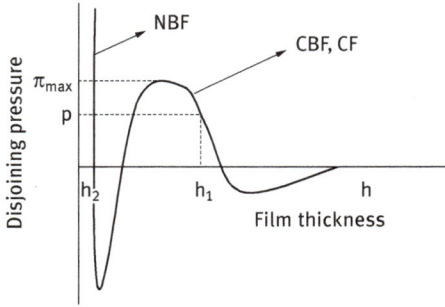

Fig. 7.1. Disjoining pressure versus film thickness showing the transition from common film (CF) to common black film (CBF) to Newton black film (NBF).

mon black film and to the Newton black film. The common black film has a thickness in the region of 30 nm, whereas the Newton black film has a thickness in the region of 4–5 nm, depending on electrolyte concentration. Several investigations were carried out to study the above transitions from common film to common black film and finally to Newton black film. For sodium dodecyl sulfate, the common black films have thicknesses ranging from 200 nm in very dilute system to about 5.4 nm. The thickness depends strongly on electrolyte concentration and the stability may be considered to be caused by the secondary minimum in the energy distance curve. In cases where the film thins further and overcomes the primary energy maximum, it will fall into the primary minimum potential energy sink, where very thin Newton black films are produced. The transition from common black films to Newton black films occurs at a critical electrolyte concentration which depends on the type of surfactant.

The rupture mechanisms of thin liquid films were considered by de Vries [15] and by Vrij and Overbeek [16]. It was assumed that thermal and mechanical disturbances (having a wave-like nature) causes film thickness fluctuations (in thin films) leading to rupture or coalescence of bubbles at a critical thickness. Vrij and Overbeek [16] carried out a theoretical analysis of the hydrodynamic interfacial force balance, and expressed the critical thickness of rupture in terms of the attractive van der Waals interaction (characterized by the Hamaker constant A), the surface or interfacial tension γ and disjoining pressure. The critical wavelength, λ_{crit}, for the perturbation to grow (assuming the disjoining pressure just exceeds the capillary pressure) was determined. Film collapse occurs when the amplitude of the fast growing perturbation was equal to the thickness of the film. The critical thickness of rupture, h_{cr}, was defined by the following equation,

$$h_{crit} = 0.267 \left(\frac{a_f A^2}{6\pi \gamma \Delta p} \right)^{1/7},$$
(7.8)

where a_f is the area of the film.

Many poorly foaming liquids with thick film lamella are easily ruptured, e.g. pure water and ethanol films (with thickness between 110 and 453 nm). Under these conditions, rupture occurs by growth of disturbances which may lead to thinner sections [17]. Rupture can also be caused by spontaneous nucleation of vapor bubbles (forming gas cavities) in the structured liquid lamella [18]. An alternative explanation for rupture of relatively thick aqueous films containing low level of surfactants is the hydrophobic attractive interaction between the surfaces that may be caused by bubble cavities [19, 20].

7.4.2.4 Stabilization by micelles (high surfactant concentrations >cmc)

At high surfactant concentrations (above the cmc), micelles of ionic or nonionic surfactants can produce organized molecular structures within the liquid film [21, 22]. This will provide an additional contribution to the disjoining pressure. Thinning of the film occurs through a stepwise drainage mechanism, referred to as stratification [23]. The ordering of surfactant micelles (or colloidal particles) in the liquid film due to the repulsive interaction provides an additional contribution to the disjoining pressure and this prevents the thinning of the liquid film.

7.4.2.5 Stabilization by lamellar liquid crystalline phases

This is particularly the case with nonionic surfactant that produce lamellar liquid crystalline structure in the film between the bubbles [24, 25]. These liquid crystals reduce film drainage as a result of the increase in viscosity of the film. In addition, the liquid crystals act as a reservoir of surfactant of the optimal composition to stabilize the foam.

7.4.2.6 Stabilization of foam films by mixed surfactants

It has been found that a combination of surfactants gives slower drainage and improved foam stability. For example, mixtures of anionic and nonionic surfactants or anionic surfactant and long chain alcohol produce much more stable films than the single components. This could be attributed to several factors. For example, addition of a nonionic surfactant to an anionic surfactant causes a reduction in the cmc of the anionic. The mixture can also produce lower surface tension compared to the individual components. The combined surfactant system also has a high surface elasticity and viscosity when compared with the single components.

7.5 Foam inhibitors

Two main types of inhibition may be distinguished: antifoamers that are added to prevent foam formation and deformers that are added to eliminate an existing foam. For example, alcohols such as octanol are effective as defoamers but ineffective as antifoamers. Since the drainage and stability of liquid films is far from being fully understood, it is very difficult at present to explain the antifoaming and foam breaking action obtained by addition of substances. This is also complicated by the fact that in many industrial processes foams are produced by unknown impurities. For these reasons, the mechanism of action of antifoamers and defoamers is far from being understood [26]. Below a summary of the various methods that can be applied for foam inhibition and foam breaking is given.

7.5.1 Chemical inhibitors that lower viscosity and increase drainage

Chemicals that reduce the bulk viscosity and increase drainage can cause a decrease in foam stability. The same applies to materials that reduce surface viscosity and elasticity (swamping the surface layer with excess compound of lower viscosity). It has been suggested that a spreading film of antifoam may simply displace the stabilizing surfactant monolayer. As the oil lens spreads and expands on the surface, the tension will be gradually reduced to a lower uniform value. This will eliminate the stabilizing effect of the interfacial tension gradients, i.e. elimination of surface elasticity. Reduction of surface viscosity and elasticity may be achieved by low molecular weight surfactants. This will reduce the coherence of the layer, e.g. by addition of small amounts of nonionic surfactants. These effects depend on the molecular structure of the added surfactant. Other materials, which are not surface active, can also destabilize the film by acting as cosolvents which reduce the surfactant concentration in the liquid layer. Unfortunately, these non-surface active materials, such as methanol or ethanol, need to be added in large quantities (> 10%).

7.5.2 Solubilized chemicals which cause antifoaming

It has been demonstrated that solubilized antifoamers such as tributyl phosphate and methyl isobutyl carbinol when added to surfactant solutions such as sodium dodecyl sulfate and sodium oleate may reduce foam formation [27]. In cases where the oils exceed the solubility limit, the emulsifier droplets of oil can have a great influence on the antifoam action. It has been claimed [27] that the oil solubilized in the micelle causes a weak defoaming action. Mixed micelle formation with extremely low concentrations of surfactant may explain the actions of insoluble fatty acid esters, alkyl phosphate esters and alkyl amines.

7.5.3 Droplets and oil lenses which cause antifoaming and defoaming

Undissolved oil droplets form in the surface of the film and this can lead to film rupture. Several examples of oils may be used: alkyl phosphates, diols, fatty acid esters and silicone oils (polydimethyl siloxane). A widely accepted mechanism for the antifoaming action of oils considers two steps: The oil drops enter the air/water interface. The oil spreads over the film causing rupture.

The antifoaming action can be rationalized [28] in terms of the balance between the entering coefficient E and the Harkins [29] spreading coefficient S which are given by the following equations:

$$E = \gamma_{W/A} + \gamma_{W/O} - \gamma_{O/A} \tag{7.9}$$

$$S = \gamma_{W/A} - \gamma_{W/O} - \gamma_{O/A}, \tag{7.10}$$

where $\gamma_{W/A}$, $\gamma_{O/A}$ and $\gamma_{W/O}$ are the macroscopic interfacial tensions of the aqueous phase, oil phase and interfacial tension of the oil/water interface respectively.

Ross and McBain [30] suggested that for efficient defoaming, the oil drop must enter the air/water interface and spread to form a duplex film at both sides of the original film. This leads to displacement of the original film, leaving an oil film which is unstable and can easily break. Ross [27] used the spreading coefficient (equation 10) as a defoaming criterion. For antifoaming both E and S should be > 0 for entry and spreading. A typical example of this type of spreading/breaking is illustrated for a hydrocarbon surfactant stabilized film. For most surfactant systems, $\gamma_{AW} = 35–45 \ \mathrm{mNm^{-1}}$ and $\gamma_{OW} = 5–10 \ \mathrm{mNm^{-1}}$ and hence for an oil to act as an antifoaming agent, γ_{OA} should be less than 25 $\mathrm{mNm^{-1}}$. This shows why low surface tension silicone oils which have surface tensions as low as 10 $\mathrm{mNm^{-1}}$ are effective.

7.5.4 Surface tension gradients (induced by antifoamers)

It has been suggested that some antifoamers act by eliminating the structure tension gradient effect in foam films by reducing the Marangoni effect. Since spreading is driven by a surface tension gradient between the spreading front and the leading edge of the spreading front, then thinning and foam rupture can occur by this surface tension gradient acting as a shear force (dragging the underlying liquid away from the source). This could be achieved by solids or liquids containing surfactant other than that stabilizing the foam. Alternatively, liquids which contain foam stabilizers at higher concentrations than that which is present in the foam may also act by this mechanism. A third possibility is the use of adsorbed vapors of surface active liquids.

7.5.5 Hydrophobic particles as antifoamers

Many solid particles with some degree of hydrophobicity were shown to cause destabilization of foams, e.g. hydrophobic silica, PTFE particles. These particles exhibit a finite contact angle when adhering to the aqueous interface. It has been suggested that many of these hydrophobic particles can deplete the stabilizing surfactant film by rapid adsorption and can cause weak spots in the film. A further mechanism was suggested based on the degree of wetting of the hydrophobic particles [31] and this led to the idea of particle bridging. For large smooth particles (large enough to touch both surfaces and with a contact angle $\theta > 90°$) dewetting can occur. Initially the Laplace pressure in the film adjacent to the particle becomes positive and causes liquid to flow away from the particle leading to enhanced drainage and formation of a "hole". In the case of $\theta < 90°$, then initially the situation is the same as for $\theta > 90°$, but as the film drains it attains a critical thickness where the film is planar and the capillary pressure becomes zero. At this point, further drainage reverses the sign of the radii of curvature causing unbalanced capillary forces which prevent drainage occurring. This can cause a stabilizing effect for certain types of particles. This means that a critical receding contact angle is required for efficient foam breaking. With particles containing rough edges, the situation is more complex, as demonstrated by Johansson and Pugh [32], using finely ground quartz particles of different size fractions. The particle surfaces were hydrophobized by methylation. These studies and others reported in the literature confirmed the importance of size, shape and hydrophobicity of the particles on foam stability.

7.5.6 Mixtures of hydrophobic particles and oils as antifoamers

The synergetic antifoaming effect of mixtures of insoluble hydrophobic particles and hydrophobic oils when dispersed in aqueous medium has been well established in the patent literature. These mixed antifoamers are very effective at very low concentrations (10–100 ppm). The hydrophobic particles could be hydrophobized silica and the oil is polydimethyl siloxane (PDMS). One possible explanation of the synergetic effect is that the spreading coefficient of PDMS oil is modified by the addition of hydrophobic particles. It has been suggested that the oil-particle mixtures form composite entities where the particles can adhere to the oil-water interface. The presence of particles adhering to the oil-water interface may facilitate the emergence of oil droplets into the air-water interface to form lenses leading to rupture of the oil-water-air film.

7.6 Assessment of foam formation and stability

A distinction must be made between foam production, measured by the height of the foam initially formed, and foam stability, the height after a given amount of time. A qualitative method for assessment of foam formation and stability is the Ross–Miles method [33]. In this test, 200 ml of a surfactant solution contained in a pipette of specified dimension with a 2.9 mm inner diameter orifice is allowed to fall 90 cm into 50 ml of the same surfactant solution contained in a cylindrical vessel maintained at a constant temperature by means of a water jacket. The height of the foam produced in the vessel is immediately recorded after all the surfactant solution has run out of the pipette (initial foam height) and then again after 5 minutes.

7.6.1 Efficiency and effectiveness of a foaming surfactant

Foam height usually increases with increasing surfactant concentration and reaches a maximum somewhat above the critical micelle concentration (cmc). Thus the cmc of a surfactant is a good measure of its efficiency as a foaming agent; the lower the cmc the more efficient the surfactant as a foamer. For a series of surfactants with the same hydrophilic head group, the longer the hydrocarbon chain length, the lower the cmc and the more efficient the surfactant as a foaming agent. The addition of electrolytes for an ionic surfactant results in a decrease of the cmc and this increases its efficiency. However, surfactants with longer hydrophobic groups are more efficient, but not necessarily more effective foaming agents. The effectiveness of a surfactant as a foaming agent appears to depend both on its effectiveness in reducing the surface tension and on the magnitude of its intermolecular cohesive forces. Since the free energy of foam formation is given by $\Delta A \gamma$ (where ΔA is the increase in the area of the liquid/gas interface as a result of foaming and γ is the surface tension), then it is clear that the lower the value of γ, the lower the amount of work required to produce the foam. The rate of attainment of surface tension reduction determines the effectiveness of a surfactant as a foaming agent. This requires measurement of surface tension as a function of time (dynamic surface tension measurements), and the faster the rate (the shorter the time required to reach the equilibrium value of surface tension) the more effective the surfactant as a foaming agent. The rate of surface tension reduction depends on its rate of adsorption at the liquid/gas interface and this depends on the diffusion coefficient of the surfactant molecule. Thus branched chain surfactants and those containing centrally located hydrophobic groups diffuse rapidly to the interface producing higher volumes of initial foam when compared with their straight chain analogues. However, to stabilize the resulting foam against collapse, the surfactant molecules must produce an interfacial film with sufficient cohesion to impart elasticity and mechanical strength to the liquid lamellae enclosing the gas bubbles in the foam. Since interchain cohesion increases with increasing the length of the hydrophobic chain, this may account for the observation that foam height often goes through a maximum with increasing the length of the chain. A too short alkyl chain produces insufficient cohesiveness, whereas a too long chain produces too much rigidity and low elasticity. In addition, a too long alkyl chain results in low solubility in water and high Krafft temperature.

References

[1]   E. Dickinson, *Introduction to Food Colloids*, Oxford University Press, 1992.

[2]   A. Scheludko, *Colloid Science*, Elsevier, Amsterdam, 1966.

[3]   A. Scheludko, *Advances Colloid Interface Sci.*, **1**, 391 (1971).

[4]   D. Exerowa and P. M. Kruglyakov, *Foam and Foam Films*, Elsevier, Amsterdam, 1997.

[5]   R. J. Pugh, *Advances Colloid and Interface Sci.*, **64**, 67 (1995).

[6]   O. Reynolds, *Phil. Trans. Royal Soc. London, Ser. A*, **177**, 157 (1886).

[7]   K. J. Mysels, *J. Phys. Chem.*, **68**, 3441 (1964).

[8]   J. Lucassen, in: *Anionic Surfactants*, E.H. Lucassen-Reynders (ed.), p. 217, Marcel Dekker, New York, 1981.

[9]   H. N. Stein, *Advances Colloid Interface Sci.*, **34**, 175 (1991).

[10]   J. T. Davies, *Proceedings of the Second International Congress of Surface Activity*, Vol. 1, J. H. Schulman (ed.), Butterworth, London, 1957.

[11]   B. V. Deryaguin and N. V Churaev, *Kolloid Zh.*, **38**, 438 (1976).

[12]   B. V. Deryaguin, *Theory of Stability of Colloids and Thin Films*, Consultant Bureau, New York, 1989.

[13]   B. V. Deryaguin and L. D. Landua, *Acta Physicochimica USSR*, **14**, 633 (1941).

[14]   E. J. Verwey and J. Th. G. Overbeek, *Theory of Stability of Lyophobic Colloids*, Elsevier, Amsterdam, 1948.

[15]   A. J. de Vries, *Disc. Faraday Soc.*, **42**, 23, 1966).

[16]   A. Vrij and J. Th. G. Overbeek, *J. Amer. Chem. Soc.*, **90**, 3074 (1968).

[17]   B. Radoev, A. Scheludko and E. Manev, *J. Colloid Interface Sci.*, **95**, 254 (1983).

[18]   V. G. Gleim, I. V. Shelomov and B. R. Shidlovskii, *J. Appl. Chem. USSR*, **32**, 1069 (1959).

[19]   R. J. Pugh and R.H. Yoon, *J. Colloid Interface Sci.*, **163**, 169 (1994).

[20]   P. M. Claesson and H. K. Christensen, *J. Phys. Chem.*, **92**, 1650 (1988).

[21]   E. S. Johnott, *Philos. Mag.*, **11**, 746 (1906).

[22]   J. Perrin, *Ann. Phys.*, **10**, 160 (1918).

[23]   L. Loeb and D. T. Wasan, *Langmuir*, **9**, 1668 (1993).

[24]   S. Frieberg, *Mol. Cryst. Liq. Cryst.*, **40**, 49 (1977).

[25]   J. E. Perez, J. E. Proust and Ter-Minassian Saraga, in: *Thin Liquid Films*, p. 70, I.B. Ivanov (ed.), Marcel Dekker, New York, 1988.

[26]   P. R. Garrett, Editor, *Defoaming*, Surfactant Science Series Vol. 45, Marcel Dekker, New York, 1993.

[27]   S. Ross and R. M Haak, *J. Phys. Chem.*, **62**, 1260 (1958).

[28]   J. V. Robinson and W. W. Woods, *J. Soc. Chem. Ind.*, **67**, 361 (1948).

[29]   W. D. Harkins, *J. Phys. Chem.*, **9**, 552 (1941).

[30]   S. Ross and McBain, *Ind. Chem. Eng.*, **36**, 570 (1944).

[31]   P. R. Garett, *J. Colloid Interface Sci.*, **69**, 107 (1979).

[32]   G. Johansson and R. J. Pugh, *Int. J. Mineral Process.*, **34**, 1 (1992).

[33]   J. Ross and G. D. Miles, *Am. Soc. For Testing Materials*, Method D1173-53, Philadelphia, PA (1953); *Oil Soap* **18**, 99 (1941).

8 Surfactants in nanoemulsions

8.1 Introduction

Nanoemulsions are transparent or translucent systems mostly covering the size range 20–200 nm [1, 2]. Nanoemulsions were also referred to as mini-emulsions [3–7]. They can be transparent, translucent or turbid depending on the droplet size, the difference in refractive index between the droplets and the dispersion medium and the disperse phase volume fraction. This can be understood from consideration of the dependence of light scattering (turbidity) on the above factors. For droplets with a radius that is less than (1/20) of the wave length of the light, the turbidity τ is given by the following equation,

$$\tau = KN_oV^2 \tag{8.1}$$

where K is an optical constant that is related to the difference in refractive index between the droplets n_p and the medium n_o, and N_o is the number of droplets each with a volume V.

It is clear from equation (8.1) that τ decreases with decreasing K, i.e. smaller $(n_p–n_o)$, decrease of N_o and decrease of V. Thus to produce a transparent nanoemulsion one has to decrease the difference between the refractive index of the droplets and the medium (i.e. try to match the two refractive indices). If such matching is not possible then one has to reduce the droplet size (by high pressure homogenization) to values below 50 nm. It is also necessary to use a nanoemulsion with low oil volume fraction (in the region of 0.2).

Unlike microemulsions (which are also transparent or translucent and thermodynamically stable, see Chapter 9) nanoemulsions are only kinetically stable. However, the long term physical stability of nanoemulsions (with no apparent flocculation or coalescence) makes them unique and they are sometimes referred to as "approaching thermodynamic stability".

The inherently high colloid stability of nanoemulsions can be well understood from a consideration of their steric stabilization (when using nonionic surfactants and/or polymers) and how this is affected by the ratio of the adsorbed layer thickness to droplet radius as will be discussed below. Unless adequately prepared (to control the droplet size distribution) and stabilized against Ostwald ripening (that occurs when the oil has some finite solubility in the continuous medium), nanoemulsions may lose their transparency with time as a result of increase in droplet size.

The attraction of nanoemulsions for applications in personal care and cosmetics as well as in health care is due to the following advantages:

1. The very small droplet size causes a large reduction in the gravity force and the Brownian motion may be sufficient for overcoming gravity. This means that no creaming or sedimentation occurs on storage.

2. The small droplet size also prevents any flocculation of the droplets. Weak flocculation is prevented and this enables the system to remain dispersed with no separation.
3. The small droplets also prevent their coalescence, since these droplets are non-deformable and hence surface fluctuations are prevented. In addition, the significant surfactant film thickness (relative to droplet radius) prevents any thinning or disruption of the liquid film between the droplets.
4. Nanoemulsions are suitable for efficient delivery of active ingredients through the skin. The large surface area of the emulsion system allows rapid penetration of actives.
5. Due to their small size, nanoemulsions can penetrate through the "rough" skin surface and this enhances penetration of actives.
6. The transparent nature of the system, their fluidity (at reasonable oil concentrations) as well as the absence of any thickeners may give them a pleasant aesthetic character and skin feel.
7. Unlike microemulsions (which require a high surfactant concentration, usually in the region of 20 % and higher), nanoemulsions can be prepared using reasonable surfactant concentration. For a 20 % O/W nanoemulsion, a surfactant concentration in the region of 5–10 % may be sufficient.
8. The small size of the droplets allows them to deposit uniformly on substrates – wetting, spreading and penetration may be also enhanced as a result of the low surface tension of the whole system and the low interfacial tension of the O/W droplets.
9. Nanoemulsions can be applied for delivery of fragrants which may be incorporated in many personal care products. This could also be applied in perfumes which are desirable to be formulated alcohol free.
10. Nanoemulsions may be applied as a substitute for liposomes and vesicles (which are much less stable) and it is possible in some cases to build lamellar liquid crystalline phases around the nanoemulsion droplets.

In this chapter, I will discuss the following topics:
1. Fundamental principles of emulsification and the role of surfactants.
2. Production of nanoemulsions using high pressure homogenizers and the phase inversion principles.
3. Theory of steric stabilization of nanoemulsions and the role of the relative ratio of adsorbed layer thickness to the droplet radius.
4. Theory of Ostwald ripening and methods of reduction of the process.
5. Examples of nanoemulsions.

8.2 Fundamental principles of emulsification

As mentioned in Chapter 6 to prepare an emulsion oil, water, surfactant and energy are needed. This can be considered from a consideration of the energy required to expand the interface, $\Delta A \gamma$ (where ΔA is the increase in interfacial area when the bulk oil with area A_1 produces a large number of droplets with area A_2; $A_2 \gg A_1$, γ is the interfacial tension). Since γ is positive, the energy to expand the interface is large and positive. This energy term cannot be compensated by the small entropy of dispersion $T\Delta S$ (which is also positive) and the total free energy of formation of an emulsion, ΔG is positive,

$$\Delta G = \Delta A \gamma - T\Delta S. \tag{8.2}$$

Thus, emulsion formation is non-spontaneous and energy is required to produce the droplets. The formation of large droplets (few μm), as is the case for macroemulsions, is fairly easy and hence high speed stirrers such as the Ultraturrax or Silverson Mixer are sufficient to produce the emulsion. In contrast, the formation of small drops (sub-micron as is the case with nanoemulsions) is difficult and this requires a large amount of surfactant and/or energy.

The high energy required for formation of nanoemulsions can be understood from a consideration of the Laplace pressure p (the difference in pressure between inside and outside the droplet,

$$p = \gamma \left(\frac{1}{R_1} + \frac{1}{R_2} \right) \tag{8.3}$$

where R_1 and R_2 are the principal radii of curvature of the drop.

For a spherical drop, $R_1 = R_2 = R$ and

$$p = \frac{2\gamma}{R}. \tag{8.4}$$

To break up a drop into smaller ones, it must be strongly deformed and this deformation increases p. Consequently, the stress needed to deform the drop is higher for a smaller drop. Since the stress is generally transmitted by the surrounding liquid via agitation, higher stresses need more vigorous agitation, hence more energy is needed to produce smaller drops [8].

Surfactants play major roles in the formation of nanoemulsions: By lowering the interfacial tension, p is reduced and hence the stress needed to break up a drop is reduced. Surfactants prevent coalescence of newly formed drops

To assess nanoemulsion formation, one usually measures the droplet size distribution using dynamic light scattering techniques (Photon Correlation Spectroscopy, PCS). In this technique, one measures the intensity fluctuation of scattered light by the droplets as they undergo Brownian motion [9]. When a light beam passes through a nanoemulsion, an oscillating dipole moment is induced in the droplets, thereby reradiating the light. Due to the random position of the droplets, the intensity of scattered light will, at any instant, appear as a random diffraction or "speckle" pattern.

As the droplets undergo Brownian motion, the random configuration of the pattern will, therefore, fluctuate such that the time taken for an intensity maximum to become a minimum, i.e. the coherence time, corresponds exactly to the time required for the droplet to move one wavelength. Using a photomultiplier of active area about the diffraction maximum, i.e. one coherence area, this intensity fluctuation can be measured. The analogue output is digitized using a digital correlator that measures the photocount (or intensity) correlation function of the scattered light. The photocount correlation function $G^{(2)}(\tau)$ is given by the equation,

$$G^{(2)}(\tau) = B(1 + \gamma^2 [g^{(1)}(\tau)]^2),\tag{8.5}$$

where τ is the correlation delay time. The correlator compares $G^{(2)}(\tau)$ for many values of τ. B is the background value to which $G^{(2)}(\tau)$ decays at long delay times. $g^{(1)}(\tau)$ is the normalized correlation function of the scattered electric field and γ is a constant (~ 1).

For monodisperse non-interacting droplets,

$$g^{(1)} = \exp(-\Gamma\tau)\tag{8.6}$$

where Γ is the decay rate or inverse coherence time that is related to the translational diffusion coefficient D by the equation,

$$\Gamma = DK^2,\tag{8.7}$$

where K is the scattering vector,

$$K = \frac{4\pi n}{\lambda_o} \sin\left(\frac{\theta}{2}\right),\tag{8.8}$$

λ is the wave length of light in vacuo, n is the refractive index of the solution and θ is the scattering angle.

The droplet radius R can be calculated from D using the Stokes–Einstein equation,

$$D = \frac{kT}{6\pi\eta_o R},\tag{8.9}$$

η_o is the viscosity of the medium.

The above analysis is valid for dilute monodisperse droplets. With many nanoemulsions, the droplets are not perfectly monodisperse (usually with a narrow size distribution) and the light scattering results are analyzed for polydispersity (the data are expressed as an average size) and a polydispersity index gives information on the deviation from the average size.

8.2.1 Methods of emulsification and the role of surfactants

As mentioned in Chapter 6, several procedures may be applied for emulsion preparation, these range from simple pipe flow (low agitation energy L), static mixers and general stirrers (low to medium energy, L–M), high speed mixers such as the Ultraturrex

(M), colloid mills and high pressure homogenizers (high energy, H), ultrasound generators (M–H). The method of preparation can be continuous (C) or batch-wise (B). With nanoemulsions a higher power density is required and this restricts the preparation of nanoemulsions to the use of high pressure homogenizers and ultrasonics.

An important parameter that describes droplet deformation is the Weber number, W_e, which gives the ratio of the external stress $G\eta$ (where G is the velocity gradient and η is the viscosity) over the Laplace pressure (see Chapter 6),

$$W_e = \frac{G\eta r}{2\gamma}. \tag{8.10}$$

Droplet deformation increases with increasing Weber number, which means that for producing small droplets one requires high stresses (high shear rates). In other words, the production of nanoemulsions costs more energy than that required to produce macroemulsions [4]. The role of surfactants on emulsion formation has been described in detail in Chapter 6 and the same principles apply to the formation of nanoemulsions. Thus, one must consider the effect of surfactants on the interfacial tension, interfacial elasticity and interfacial tension gradients.

8.3 Preparation of nanoemulsions

Two methods may be applied for the preparation of nanoemulsions (covering the droplet radius size range 50–200 nm). Use of high pressure homogenizers (aided by appropriate choice of surfactants and cosurfactants) or application of the phase inversion concepts.

8.3.1 Use of high pressure homogenizers

The production of small droplets (submicron) requires application of high energy and the process of emulsification is generally inefficient. Simple calculations show that the mechanical energy required for emulsification exceeds the interfacial energy by several orders of magnitude. For example to produce an emulsion at $\phi = 0.1$ with a $d_{32} = 0.6$ μm, using a surfactant that gives an interfacial tension $\gamma = 10$ mNm^{-1}, the net increase in surface free energy is $A\gamma = 6\phi\gamma/d_{32} = 10^4$ Jm^{-3}. The mechanical energy required in a homogenizer is 10^7 Jm^{-3}, i.e. an efficiency of 0.1 % – the rest of the energy (99.9 %) is dissipated as heat [10].

The intensity of the process or the effectiveness in making small droplets is often governed by the net power density ($\epsilon(t)$),

$$p = \epsilon(t)dt, \tag{8.11}$$

where t is the time during which emulsification occurs.

Break up of droplets will only occur at high ϵ values, which means that the energy dissipated at low ϵ levels is wasted. Batch processes are generally less efficient than continuous processes. This shows why with a stirrer in a large vessel, most of the energy applied at low intensity is dissipated as heat. In a homogenizer, p is simply equal to the homogenizer pressure.

Several procedures may be applied to enhance the efficiency of emulsification when producing nanoemulsions: One should optimize the efficiency of agitation by increasing ϵ and decreasing dissipation time. The emulsion is preferably prepared at high volume faction of the disperse phase and diluted afterwards. However, very high ϕ values may result in coalescence during emulsification. It is preferable to use high surfactant concentration, whereby creating a smaller γ_{eff} and possibly diminishing recoalescence. Using surfactant mixtures that show more reduction in γ than the individual components is also essential. If possible one should dissolve the surfactant in the disperse phase rather than the continuous phase; this often leads to smaller droplets. It may be useful to emulsify in steps of increasing intensity, particularly with emulsions having highly viscous disperse phase.

8.3.2 Phase inversion principle methods (low energy emulsification)

8.3.2.1 Phase inversion composition (PIC) method

A study of the phase behavior of water/oil/surfactant systems demonstrated that emulsification can be achieved by three different low energy emulsification methods, as schematically shown in Fig. 8.1:

(A) stepwise addition of oil to a water surfactant mixture;
(B) stepwise addition of water to a solution of the surfactant in oil;
(C) mixing all the components in the final composition, pre-equilibrating the samples prior to emulsification.

In these studies [5], the system water/Brij 30 (polyoxyethlene lauryl ether with an average of 4 moles of ethylene oxide)/decane was chosen as a model to obtain O/W emulsions. The results showed that nanoemulsions with droplet sizes of the order of 50 nm were formed only when water was added to mixtures of surfactant and oil (method) whereby inversion from W/O emulsion to O/W nanoemulsion occurred.

The PIC method makes use of the chemical energy released during the emulsification process as a consequence of a change in the spontaneous curvature of surfactant molecules, from negative to positive (obtaining O/W nanoemulsions) or from positive to negative (obtaining W/O nanoemulsions). In the PIC method, the change in curvature is induced by the progressive addition of the intended continuous phase which may be pure water or oil. At the inversion point, the surfactant film adopts a planer configuration and the interfacial tension approaches zero. This explains why nanoemulsions are produced at and above the inversion point.

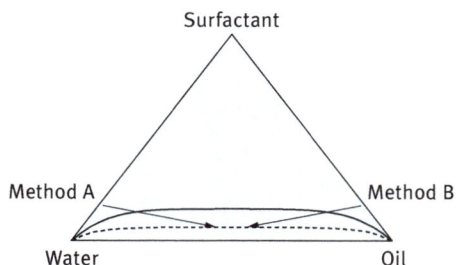

Fig. 8.1. Schematic representation of the experimental path in two emulsification methods: method A, addition of decane to water/surfactant mixture; method B, addition of water to decane/Brij 30 solutions.

8.3.2.2 Phase inversion temperature (PIT) method

This method has been demonstrated by Shinoda and coworkers [11, 12] when using nonionic surfactants of the ethoxylate type. These surfactants are highly dependent on temperature, becoming lipophilic with increasing temperature due to the dehydration of the polyethyleneoxide chain. When an O/W emulsion is prepared using a nonionic surfactant of the ethoxylate type and is heated, then at a critical temperature (the PIT), the emulsion inverts to a W/O emulsion. At the PIT the hydrophilic and lipophilic components of the surfactant are exactly balanced and the PIT is sometimes referred to as the HLB temperature. This is illustrated in Fig. 8.2 which shows the variation of interfacial tension γ of n-octane/water system with temperature for a series of alcohol ethoxylates. A clear minimum in γ is observed at a critical temperature (PIT) that depends on the length of the alkyl chain and number of ethylene oxide units. At the PIT the droplet size reaches a minimum and the interfacial tension also reaches a minimum [13, 14]. However, the small droplets are unstable and they coalesce very rapidly. By rapid cooling of the emulsion that is prepared at a temperature near the PIT, very stable and small emulsion droplets could be produced. Thus by preparing the emulsion at a temperature 2–4°C below the PIT (near the minimum in γ) followed by rapid cooling of the system, nanoemulsions may be produced.

The minimum in γ can be explained in terms of the change in curvature H of the interfacial region, as the system changes from O/W to W/O. For O/W system and normal micelles, the monolayer curves towards the oil and H is given a positive value. For a W/O emulsions and inverse micelles, the monolayer curves towards the water and H is assigned a negative value. At the inversion point (HLB temperature) H becomes zero and γ reaches a minimum.

8.4 Steric stabilization and the role of the adsorbed layer thickness

Since most nanoemulsions are prepared using nonionic and/or polymeric surfactants, it is necessary to consider the interaction forces between droplets containing adsorbed

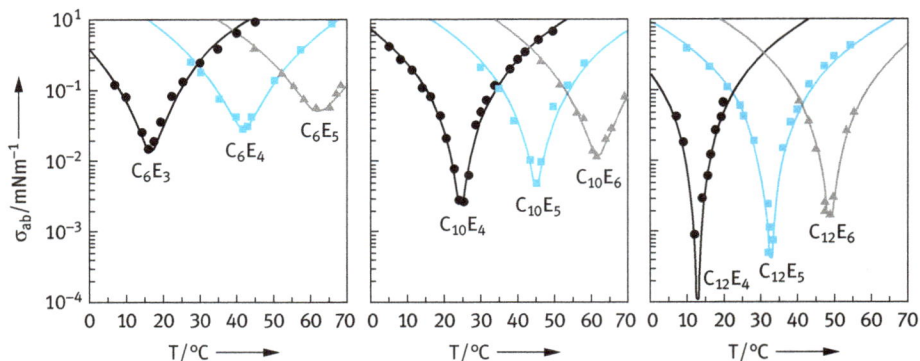

Fig. 8.2. Interfacial tensions of n-octane against water in the presence of various C_nE_m surfactants above the cmc as a function of temperature.

layers (steric stabilization). This was described in detail in Chapter 6 and only a summary is given here [15, 16].

When two droplets each containing an adsorbed layer of thickness δ approach to a distance of separation h, whereby h becomes less than 2δ, repulsion occurs as result of two main effects:

(i) Unfavorable mixing of the stabilizing chains A of the adsorbed layers, when these are in good solvent conditions. This is referred to as the mixing (osmotic) interaction, G_{mix}, and is given by the following expression,

$$\frac{G_{mix}}{kT} = \frac{4\pi}{3V_1}\phi_2^2\left(\frac{1}{2} - \chi\right)\left(3a + 2\delta + \frac{h}{2}\right) \tag{8.12}$$

where k is the Boltzmann constant, T is the absolute temperature, V_1 is the molar volume of the solvent, ϕ_2 is the volume fraction of the polymer (the A chains) in the adsorbed layer and χ is the Flory–Huggins (polymer-solvent interaction) parameter.

It can be seen that G_{mix} depends on three main parameters: the volume fraction of the A chains in the adsorbed layer (the more dense the layer is, the higher the value of G_{mix}); the Flory–Huggins interaction parameter χ (for G_{mix} to remain positive, i.e. repulsive, χ should be lower than 1/2); and the adsorbed layer thickness δ.

(ii) Reduction in configurational entropy of the chains on significant overlap – this is referred to as elastic (entropic) interaction and is given by the expression,

$$G_{el} = 2v_2 \ln\left[\frac{\Omega(h)}{\Omega(\infty)}\right], \tag{8.13}$$

where v_2 is the number of chains per unit area, $\Omega(h)$ is the configurational entropy of the chains at a separation distance h and $\Omega(\infty)$ is the configurational entropy at infinite distance of separation.

Combination of G_{mix}, G_{el} and the van der Waals attraction G_A gives the total energy of interaction G_T,

$$G_T = G_{mix}, +, G_{el} + G_A . \qquad (8.14)$$

Fig. 8.3 gives a schematic representation of the variation of G_{mix}, G_{el}, G_A and G_T with h. As can be seen from Fig. 8.3, G_{mix} increases very rapidly with decreasing h as soon as $h < 2\delta$, G_{el} increases very rapidly with decreasing h when $h < \delta$. G_T shows one minimum, G_{min}, and it increases very rapidly with decreasing h when $h < 2\delta$.

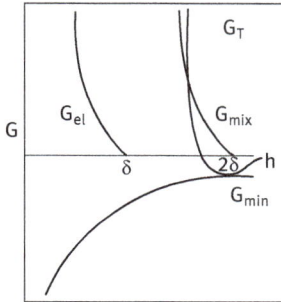

Fig. 8.3. Energy-distance curves for sterically stabilized nanoemulsions.

The magnitude of G_{min} depends on the following parameters: the particle radius R, the Hamaker constant A, and the adsorbed layer thickness δ. As an illustration, Fig. 8.4 shows the variation of G_T with h at various ratios of δ/R.

It can be seen from Fig. 8.4 that the depth of the minimum decrease with increasing δ/R. This is the basis of the high kinetic stability of nanoemulsions. With nanoemulsions having a radius in the region of 50 nm and an adsorbed layer thickness of say 10 nm, the value of δ/R is 0.2. This high value (when compared with the situation with macroemulsions where δ/R is at least an order of magnitude lower) results in a very shallow minimum (which could be less than kT).

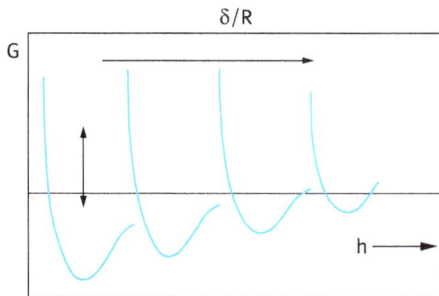

Fig. 8.4. Importance of adsorbed layer thickness to particle size ratio (d/R).

The above situation results in very high stability with no flocculation (weak or strong). In addition, the very small size of the droplets and the dense adsorbed layers ensures lack of deformation of the interface, lack of thinning and disruption of the liquid film between the droplets and hence coalescence is also prevented.

The only instability problem with nanoemulsions is Ostwald ripening which is discussed below.

8.5 Ostwald Ripening

One of the main problems with nanoemulsions is Ostwald ripening which results from the difference in solubility between small and large droplets. The difference in chemical potential of dispersed phase droplets between different sized droplets was given by Lord Kelvin [17],

$$c(r) = c(\infty) \exp\left(\frac{2\gamma V_m}{r\,RT}\right), \tag{8.15}$$

where $S(r)$ is the solubility surrounding a particle of radius r, $S(\infty)$ is the bulk phase solubility (for an infinitely large droplets) and V_m is the molar volume of the dispersed phase.

The quantity $(2\gamma V_m/RT)$ is termed the characteristic length. It has an order of ~ 1 nm or less, indicating that the difference in solubility of a 1 μm droplet is of the order of 0.1 % or less.

Theoretically, Ostwald ripening should lead to condensation of all droplets into a single drop (i.e. phase separation). This does not occur in practice since the rate of growth decreases with increasing droplet size.

For two droplets of radii r_1 and r_2 (where $r_1 < r_2$),

$$\left(\frac{RT}{V_m}\right) \ln\left[\frac{c(r_1)}{c(r_2)}\right] = 2\gamma\left(\frac{1}{r_1} - \frac{1}{r_2}\right). \tag{8.16}$$

Equation (8.16) shows that the larger the difference between r_1 and r_2, the higher the rate of Ostwald ripening.

Ostwald ripening can be quantitatively assessed from plots of the cube of the radius versus time t (the Lifshitz–Slesov–Wagner, LSW, Theory) [18, 19],

$$r^3 = \frac{8}{9}\left[\frac{c(\infty)\gamma V_m D}{\rho RT}\right] t, \tag{8.17}$$

where D is the diffusion coefficient of the disperse phase in the continuous phase and ρ is the density of the disperse phase.

Several methods may be applied to reduce Ostwald ripening [20–22]:
1. Addition of a second disperse phase component which is insoluble in the continuous phase (e.g. squalene). In this case significant partitioning between different droplets occurs, with the component having low solubility in the continuous

phase expected to be concentrated in the smaller droplets. During Ostwald ripening in two component disperse phase system, equilibrium is established when the difference in chemical potential between different size droplets (which results from curvature effects) is balanced by the difference in chemical potential resulting from partitioning of the two components. If the secondary component has zero solubility in the continuous phase, the size distribution will not deviate from the initial one (the growth rate is equal to zero). In the case of limited solubility of the secondary component, the distribution is the same as governed by equation (8.17), i.e. a mixture growth rate is obtained which is still lower than that of the more soluble component.

2. Modification of the interfacial film at the O/W interface: According to equation (8.16) reduction in γ results in reduction of Ostwald ripening. However, this alone is not sufficient since one has to reduce γ by several orders of magnitude. Walstra [23] suggested that by using surfactants which are strongly adsorbed at the O/W interface (i.e. polymeric surfactants) and which do not desorb during ripening, the rate could be significantly reduced. An increase in the surface dilational modulus and decrease in γ would be observed for the shrinking drops. The difference in γ between the droplets would balance the difference in capillary pressure (i.e. curvature effects). To achieve this above effect it is useful to use A–B–A block copolymers that are soluble in the oil phase and insoluble in the continuous phase, e.g. a tribloc of PHS-PEO-PHS, where PHS refers to polyhydroxystearic acid and PEO refers to polyethylene oxide. The polymeric surfactant should enhance the lowering of γ by the emulsifier. In other words, the emulsifier and the polymeric surfactant should show synergy in lowering γ.

8.6 Examples of nanoemulsions

Several experiments were carried out to investigate the methods of preparation of nanoemulsions and their stability [24]. The first method applied the PIT principle for preparation of nanoemulsions. Experiments were carried out using hexadecane and isohexadecane (Arlamol HD) as the oil phase and $C_{12}H_{25}-O(CH_2-CH_2-O)_4H$ ($C_{12}EO_4$) as the nonionic emulsifier. The HLB temperature was determined using conductivity measurements, whereby 10^{-2} mol dm^{-3} NaCl was added to the aqueous phase (to increase the sensitivity of the conductibility measurements). The concentration of NaCl was low and hence it had little effect on the phase behavior.

Fig. 8.5 shows the variation of conductivity versus temperature for 20 % O/W emulsions at different surfactant concentrations. It can be seen that there is a sharp decrease in conductivity at the PIT or HLB temperature of the system.

The HLB temperature decreases with increasing surfactant concentration. This could be due to the excess nonionic surfactant remaining in the continuous phase.

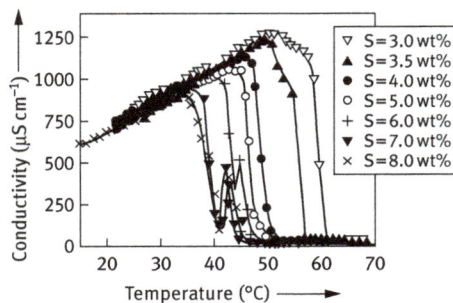

Fig. 8.5. Variation of conductivity with temperature for a 20 : 80 hexadecane-water emulsion at various $C_{12}EO_4$ surfactant (S) concentrations.

Nanoemulsions were prepared by rapid cooling of the system to 25°C. The droplet diameter was determined using photon correlation spectroscopy (PCS). The results are summarized in Table 8.1, which shows the exact composition of the emulsions, HLB temperature, z-average radius and polydispersity index.

Table 8.1. Composition, HLB temperature (T_{HLB}), droplet radius r and polydispersity index (Poly.index) for the system water-$C_{12}EO_4$-hexadecane at 25°C.

Surfactant (wt %)	Water (wt %)	Oil/Water T_{HLB} (°C)	r/nm	Poly.	index
2.0	78.0	20.4/79.6	—	320	1.00
3.0	77.0	20.6/79.4	57.0	82	0.41
3.5	76.5	20.7/79.3	54.0	69	0.30
4.0	76.0	20.8/79.2	49.0	66	0.17
5.0	75.0	21.2/78.9	46.8	48	0.09
6.0	74.0	21.3/78.7	45.6	34	0.12
7.0	73.0	21.5/78.5	40.9	30	0.07
8.0	72.0	21.7/78.3	40.8	26	0.08

O/W nanoemulsions with droplet radii in the range 26–66 nm could be obtained at surfactant concentrations between 3 and 8 %. The nanoemulsion droplet size and polydispersity index decreases with increasing surfactant concentration

The decrease in droplet size with increasing surfactant concentration is due to the increase in surfactant interfacial area and decrease in interfacial tension, γ. As mentioned above, γ reaches a minimum at the HLB temperature. Therefore, the minimum in interfacial tension occurs at lower temperature as the surfactant concentration increases. This temperature becomes closer to the cooling temperature as the surfactant concentration increases and this results in smaller droplet sizes.

All nanoemulsions showed an increase in droplet size with time, as a result of Ostwald ripening. Fig. 8.6 shows plots of r^3 versus time for all the nanoemulsions studied. The slope of the lines gives the rate of Ostwald ripening ω ($m^3 s^{-1}$) and this showed an

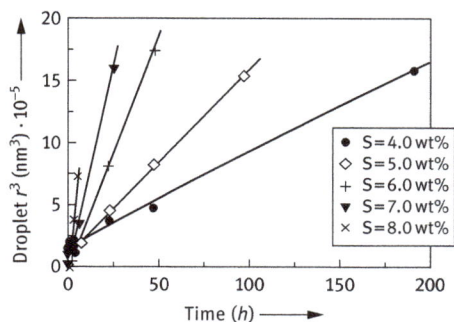

Fig. 8.6. r^3 versus time at 25°C for nanoemulsions prepared using water-$C_{12}EO_4$-hexadecane.

increase from 2×10^{-27} to 39.7×10^{-27} m^3 s^{-1} as the surfactant concentration is increased from 4 to 8 wt %. This increase could be due to three main factors:

1. decrease in droplet size increases Brownian diffusion and this enhances the rate;
2. presence of micelles, which increase with increasing surfactant concentration. This has the effect of increasing the solubilization of the oil into the core of the micelles;
3. partition of surfactant molecules between the oil and the aqueous phases. With higher surfactant concentrations, the molecules with shorter EO chains (lower HLB number) may preferentially accumulate at the O/W interface and this may result in reduction of the Gibbs elasticity, which in turn results in an increase in the Ostwald ripening rate.

The results with isohexadecane are summarized in Table 8.2.

Table 8.2. Composition, HLB temperature (T_{HLB}), droplet radius r and polydispersity index (pol.) at 25° for emulsions in the system water-$C_{12}EO_4$-isohexadecane.

Surfactant (wt %)	Water (wt %)	O/W	T_{HLB}/°C	r/nm	pol.
2.0	78.0	20.4/79.6	—	97	0.50
3.0	77.0	20.6/79.4	51.3	80	0.13
4.0	76.0	20.8/79.2	43.0	65	0.06
5.0	75.0	21.1/78.9	38.8	43	0.07
6.0	74.0	21.3/78.7	36.7	33	0.05
7.0	73.0	21.3/78.7	33.4	29	0.06
8.0	72.0	21.7/78.3	32.7	27	0.12

As with the hexadecane system, the droplet size and polydispersity index decreased with increasing surfactant concentration. Nanoemulsions with droplet radii of 25–80 nm were obtained at 3–8 % surfactant concentration. It should be noted, however, that nanoemulsions could be produced at lower surfactant concentration when using isohexadecane, when compared with the results obtained with hexadecane. This could be attributed to the higher solubility of the isohexadecane (a branched hydrocarbon), the lower HLB temperature and the lower interfacial tension.

The stability of the nanoemulsions prepared using isohexadecane was assessed by following the droplet size as a function of time. Plots of r^3 versus time for four surfactant concentrations (3, 4, 5 and 6 wt %) are shown in Fig. 8.7. The results show an increase in Ostwald ripening rate as the surfactant concentration is increased from 3 to 6 % (the rate increased from 4.1×10^{-27} to 50.7×10^{-27} m^{3}s^{-1}). The nanoemulsions prepared using 7 wt % surfactant were so unstable that they showed significant creaming after 8 hours. However, when the surfactant concentration was increased to 8 wt %, a very stable nanoemulsion could be produced with no apparent increase in droplet size over several months. This unexpected stability was attributed to the phase behavior at such surfactant concentrations. The sample containing 8 wt % surfactant showed birefringence to shear when observed under polarized light. It seems that the ratio between the phases ($W_m + L_\alpha + O$) may play a key factor in nanoemulsion stability.

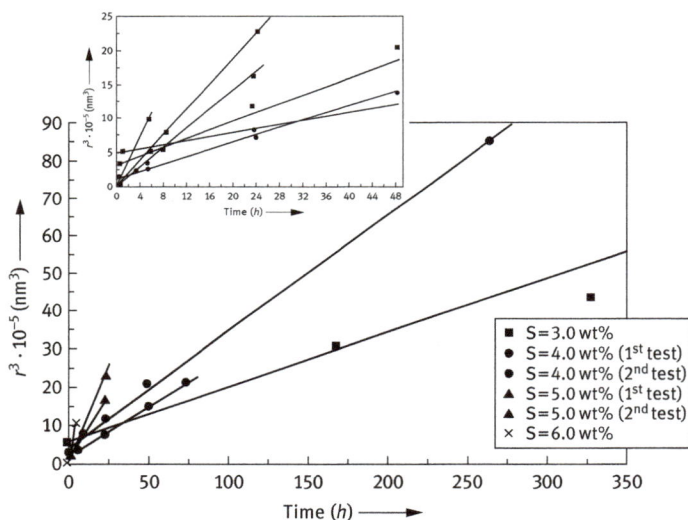

Fig. 8.7. r^3 versus time at 25°C for the system water-$C_{12}EO_4$-isohexadecane at various surfactant concentrations; O/W ratio 20/80.

Attempts were made to prepare nanoemulsions at higher O/W ratios (hexadecane being the oil phase), while keeping the surfactant concentration constant at 4 wt %. When the oil content was increased to 40 and 50 %, the droplet radius increased to 188 and 297 nm respectively. In addition, the polydispersity index also increased to 0.95. These systems become so unstable that they showed creaming within a few hours.

This is not surprising, since the surfactant concentration is not sufficient to produce the nanoemulsion droplets with high surface area. Similar results were obtained with isohexadecane. However, nanoemulsions could be produced using 30/70 O/W ratio (droplet size being 81 nm), but with high polydispersity index (0.28). The nanoemulsions showed significant Ostwald ripening.

The effect of changing the alkyl chain length and branching was investigated using decane, dodecane, tetradecane, hexadecane and isohexadecane. Plots of r^3 versus time are shown in Fig. 8.8 for 20/80 O/W ratio and surfactant concentration of 4 wt %. As expected, by reducing the oil solubility from decane to hexadecane, the rate of Ostwald ripening decreases. The branched oil isohexadecane also shows a higher Ostwald ripening rate when compared with hexadecane. A summary of the results is shown in Table 8.3 which also shows the solubility of the oil $C(\infty)$.

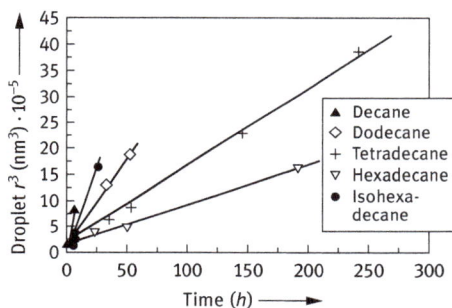

Fig. 8.8. r^3 versus time at 25 °C for nanoemulsions (O/W ratio 20/80) with hydrocarbons of various alkyl chain lengths. System water-$C_{12}EO_4$-hydrocarbon (4 wt % surfactant).

Table 8.3. HLB temperature (T_{HLB}), droplet radius r, Ostwald ripening rate ω and oil solubility for nanoemulsions prepared using hydrocarbons with different alkyl chain length.

Oil	T_{HLB}/°C	r/nm	$\omega.10^{27}$ m^{3}s^{-1}	$C\infty$) ml.ml^{-1}
Decane	38.5	59	20.9	710.0
Dodecane	45.5	62	9.3	52.0
Tetradecane	49.5	64	4.0	3.7
Heaxadecane	49.8	66	2.3	0.3
Isohexadecane	43.0	60	8.0	—

As expected from the Ostwald ripening theory (LSW theory, equation (8.17), the rate of Ostwald ripening decreases as the oil solubility decreases. Isohexadecan has a rate of Ostwald ripening similar to that of dodecane.

As discussed before, one would expect that the Ostwald ripening of any given oil should decrease on incorporation of a second oil with much lower solubility. To test this hypothesis, nanoemulsions were made using hexadecane or isohexadecane to which various proportions of a less soluble oil, namely squalene, was added. The results using hexadecane did show a significant decrease in stability on addition of 10 % squalene. This was thought to be due to coalescence rather than an increased Ostwald ripening rate. In some cases, addition of a hydrocarbon with a long alkyl chain can induce instability as a result of changes in the adsorption and conformation of the surfactant at the O/W interface. In contrast to the results obtained with hexadecane, addition of squalene to the O/W nanoemulsion system based on isohexadecane showed a systematic decrease in Ostwald ripening rate as the squalene content was increased. Addition of squalene up to 20 % based on the oil phase showed a systematic reduction in the rate (from 8.0×10^{27} to 4.1×10^{27} $m^3 s^{-1}$). It should be noted that when squalene alone was used as the oil phase, the system was very unstable and it showed creaming within 1 hour. This shows that the surfactant used is not suitable for emulsification of squalene.

The effect of HLB number on nanoemulsion formation and stability was investigated by using mixtures of $C_{12}EO_4$ (HLB = 9.7) and $C_{12}EO_4$ (HLB = 11.7). Two surfactant concentrations (4 and 8 wt %) were used and the O/W ratio was kept at 20/80. The droplet radius remained virtually constant in the HLB range 9.7–11.0, after which there is a gradual increase in droplet radius with increasing HLB number of the surfactant mixture. All nanoemulsions showed an increase in droplet radius with time, except for the sample prepared at 8 wt % surfactant with an HLB number of 9.7 (100 % $C_{12}EO_4$). Fig. 8.9 shows the variation of Ostwald ripening rate constant ω with HLB number of surfactant. The rate seems to decrease with increasing surfactant HLB number and when the latter is > 10.5, the rate reaches a low value ($< 4 \times 10^{-27}$ $m^3 s^{-1}$).

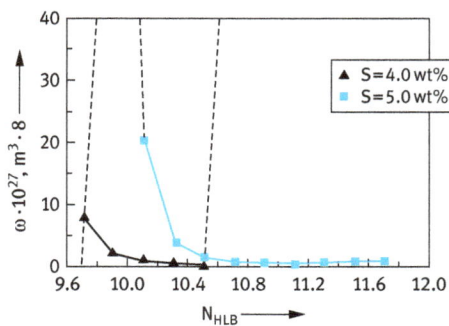

Fig. 8.9. ω versus HLB number in the systems water-$C_{12}EO_4$-$C_{12}EO_6$-isohexadecane at two surfactant concentrations.

As discussed above, with the incorporation of an oil soluble polymeric surfactant that adsorbs strongly at the O/W interface, one would expect a reduction in the Ostwald ripening rate. To test this hypothesis, an A–B–A block copolymer of polyhydroxystearic acid (PHS, the A chains) and polyethylene oxide (PEO, the B chain) PHS–PEO–PHS (Arlacel P135) was incorporated in the oil phase at low concentrations (the ratio of surfactant to Arlacel was varied between 99 : 1 to 92 : 8). For the hexadecane system, the Ostwald ripening rate showed a decrease with the addition of Arlacel P135 surfactant at ratios lower than 94:6. Similar results were obtained using isohexadecane. However, at higher polymeric surfactant concentrations, the nanoemulsion became unstable.

As mentioned above, the nanoemulsions prepared using the PIT method are relatively polydisperse and they generally give higher Ostwald ripening rates when compared to nanoemulsions prepared using high pressure homogenization techniques. To test this hypothesis, several nanoemulsions were prepared using a Microfluidiser (that can apply pressures in the range 5,000–15,000 psi or 350–1,000 bar). Using an oil:surfactant ratio of 4 : 8 and O/W ratios of 20 : 80 and 50 : 50, emulsions were prepared first using the Ultturrax followed by high pressure homogenization (ranging from 1,500 to 15,000 psi) The best results were obtained using a pressure of 15,000 psi (one cycle of homogenization). As expected, the nanoemulsions prepared using high pressure homogenization showed a lower Ostwald ripening rate when compared to the systems prepared using the PIT method. This is illustrated in Fig. 8.10 which shows plots of r^3 versus time for the two systems.

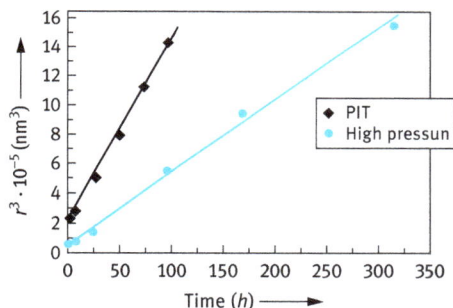

Fig. 8.10. r^3 versus time for nanoemulsion systems prepared using the PIT and Microfluidiser. 20 : 80 O/W and 4 wt % surfactant.

References

[1]   H. Nakajima, S. Tomomossa and M. Okabe, *First Emulsion Conference*, Paris, 1993.
[2]   H. Nakajima, in: *Industrial Applications of Microemulsions*, C. Solans and H. Konieda (eds.), Marcel Dekker, New York, 1997.
[3]   J. Ugelstadt, M. S. El-Aassar and J. W. Vanderhoff, *J. Polym. Sci.*, **11**, 503 (1973).
[4]   M. El-Aasser, in: *Polymeric Dispersions*, J. M. Asua (ed.), Kluwer Academic Publications, The Netherlands, 1997.
[5]   A. Forgiarini, J. Esquena, J. Gonzalez and C. Solans, *Prog. Colloid Polym. Sci.*, **115**, 36 (2000).
[6]   K. Shinoda and H. Kunieda, in: *Encyclopedia of Emulsion Technology*, P. Becher (ed.), Marcel Dekker, New York, 1983.
[7]   S. Benita and M. Y. Levy, *J. Pharm. Sci.*, **82**, 1069 (1993).

[8]   P. Walstra, in: *Encyclopedia of Emulsion Technology*, P. Becher (ed.), Marcel Dekker, New York, 1983.

[9]   P. N. Pusey, in: *Industrial Polymers: Characterisation by Molecular Weights*, J. H. S. Green and R. Dietz (eds.), Transcripta Books, London, 1973.

[10] P. Walstra and P. E. A. Smoulders, in: *Modern Aspects of Emulsion Science*, B. P. Binks (ed.), The Royal Society of Chemistry, Cambridge, 1998.

[11] K. Shinoda and H. Saito, *J. Colloid Interface Sci.*, **30**, 258 (1969).

[12] K. Shinoda and H. Saito, *J. Colloid Interface Sci.*, **26**, 70 (1968).

[13] B. W. Brooks, H. N. Richmond and M. Zerfa, in: *Modern Aspects of Emulsion Science*, B. P. Binks (ed.), Royal Society of Chemistry Publication, Cambridge, 1998.

[14] T. Sottman and R. Strey, *J. Chem. Phys.*, **108**, 8606 (1997).

[15] D. H. Napper, *Polymeric Stabilisation of Colloidal Dispersions*, Academic Press, London, 1983.

[16] Th. F. Tadros, Polymer Adsorption and Colloid Stability, in: *The Effect of Polymers on Dispersion Properties*, Th. F. Tadros (ed.), Academic Press, London, 1982.

[17] W. Thompson (Lord Kelvin), *Phil. Mag.*, **42**, 448 (1871).

[18] I. M. Lifshitz and V. V. Slesov, *Sov. Phys. JETP*, **35**, 331 (1959).

[19] C. Wagner, *Z. Electrochem.*, **35**, 581 (1961).

[20] A. S. Kabalnov and E. D. Shchukin, *Adv. Colloid Interface Sci.*, **38**, 69 (1992).

[21] A. S. Kabalnov, *Langmuir*, **10**, 680 (1994).

[22] J. G. Weers, in: *Modern Aspects of Emulsion Science*, B. P. Binks (ed.), Royal Society of Chemistry Publication, Cambridge, 1998.

[23] P. Walstra, *Chem. Eng. Sci.*, **48**, 333 (1993).

[24] P. Izquierdo, Studies on Nano-Emulsion Formation and Stability, Thesis, University of Barcelona, Spain (2002).

9 Surfactants in microemulsions

9.1 Introduction

Microemulsions are a special class of "dispersions" (transparent or translucent) that are better described as "swollen micelles". The term microemulsion was first introduced by Hoar and Schulman [1, 2] who discovered that by titration of a milky emulsion (stabilized by soap such as potassium oleate) with a medium chain alcohol such as pentanol or hexanol, a transparent or translucent system is produced. The final transparent or translucent system is a W/O microemulsion.

A convenient way to describe microemulsions is to compare them with micelles. The latter, which are thermodynamically stable, may consist of spherical units with a radius that is usually less than 5 nm. Two types of micelles may be considered: normal micelles, with the hydrocarbon tails forming the core and the polar head groups in contact with the aqueous medium, and reverse micelles (formed in nonpolar media), with a water core containing the polar head groups and the hydrocarbon tails now in contact with the oil. The normal micelles can solubilize oil in the hydrocarbon core forming O/W microemulsions, whereas the reverse micelles can solubilize water forming a W/O microemulsion.

A schematic representation of these systems is shown in Fig. 9.1. A rough guide to the dimensions of micelles, micellar solutions and macroemulsions is as follows: micelles, R < 5 nm (they scatter little light and are transparent); macroemulsions, R > 50 nm (opaque and milky); micellar solutions or microemulsions, 5–50 nm (transparent, 5–10 nm, translucent 10–50 nm).

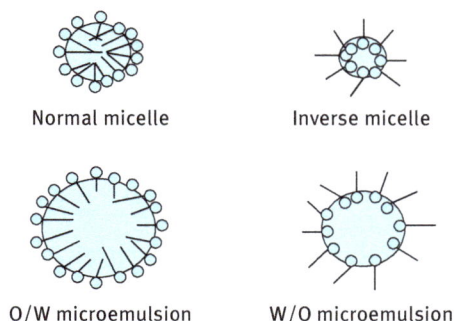

Normal micelle   Inverse micelle

O/W microemulsion   W/O microemulsion

Fig. 9.1. Schematic representation of microemulsions.

The classification of microemulsions based on size is not adequate. Whether a system is transparent or translucent depends not only on the size but also on the difference in refractive index between the oil and the water phases. A microemulsion with small size (in the region of 10 nm) may appear translucent if the difference in refractive index between the oil and the water is large (note that the intensity of light scattered depends on the size and an optical constant that is given by the difference in refractive index between oil and water). Relatively large sized microemulsion droplets (in the region of 50 nm) may appear transparent if the refractive index difference is very small. The best definition of microemulsions is based on the application of thermodynamics as is discussed below.

9.2 Thermodynamic definition of microemulsions

This can be obtained from a consideration of the energy and entropy terms for formation of microemulsions (schematically represented in Fig. 9.2) which shows schematically the process of formation of microemulsion from a bulk oil phase (for O/W microemulsion) or bulk water phase (for a W/O microemulsion).

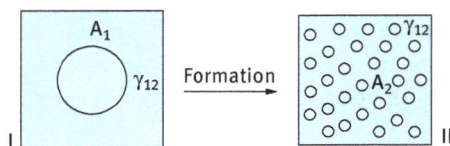

Fig. 9.2. Schematic representation of microemulsion formation.

A_1 is the surface area of the bulk oil phase and A_2 is the total surface area of all the microemulsion droplets. γ_{12} is the O/W interfacial tension.

The increase in surface area when going from state I to state II is $\Delta A \ (= A_2 - A_1)$ and the surface energy increase is equal to $\Delta A \gamma_{12}$. The increase in entropy when going from state I to state II is $T \Delta S^{conf}$ (note that state II has higher entropy since a large number of droplets can arrange themselves in several ways, whereas state I with one oil drop has much lower entropy).

According to the second law of thermodynamics, the free energy of formation of microemulsions ΔG_m is given by the following expression,

$$\Delta G_m = \Delta A \gamma_{12} - T \Delta S^{conf}. \tag{9.1}$$

With macroemulsions and nanoemulsions $\Delta A \gamma_{12} \gg T \Delta S^{conf}$ and $\Delta G_m > 0$. The system is non-spontaneous (it requires energy for formation of the emulsion drops) and it is thermodynamically unstable. With microemulsions, $\Delta A \gamma_{12} \leq T \Delta S^{conf}$ (this is due to the ultralow interfacial tension accompanying microemulsion formation) and $\Delta G_m \leq 0$. The system is produced spontaneously and it is thermodynamically stable.

The above analysis shows the contrast between emulsions, nanoemulsions and microemulsions: with emulsions and nanoemulsions, an increase in the mechanical energy and an increase in surfactant concentration usually result in the formation of smaller droplets which become kinetically more stable. With microemulsions, neither mechanical energy nor increasing surfactant concentration can result in its formation. The latter is based on a specific combination of surfactants and specific interaction with the oil and the water phases and the system is produced at optimum composition.

Thus, microemulsions have nothing in common with macroemulsions and nanoemulsions and in many cases it is better to describe the system as "swollen micelles". The best definition of microemulsions is as follows [3]: "System of Water + Oil + Amphiphile that is a single Optically Isotropic and Thermodynamically Stable Liquid Solution". Amphiphiles refer to any molecule that consists of hydrophobic and hydrophilic portions, e.g. surfactants, alcohols, etc.

The driving force for microemulsion formation is the low interfacial energy which is overcompensated by the negative entropy of dispersion term. The low (ultralow) interfacial tension is produced in most cases by combination of two molecules, referred to as the surfactant and cosurfactant (e.g. medium chain alcohol).

9.3 Description of microemulsions using phase diagrams

Consider the phase diagram of a three component system of water, ionic surfactant and medium chain alcohol as described in Fig. 9.3.

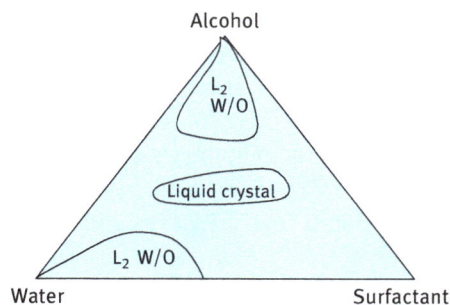

Fig. 9.3. Schematic representation of three-component phase diagram.

At the water corner and at low alcohol concentration, normal micelles (L_1) are formed since in this case there are more surfactant than alcohol molecules. At the alcohol (cosurfactant) corner, inverse micelles (L_2) are formed, since in this region there are more alcohol than surfactant molecules. These L_1 and L_2 are not in equilibrium but are separated by a liquid crystalline region (lamellar structure with equal number of surfactant and alcohol molecules). The L_1 region may be considered as an O/W microemulsion, whereas the L_2 may be considered as a W/O microemulsion.

Addition of a small amount of oil, miscible with the cosurfactant but not with the surfactant, and water changes the phase diagram only slightly. The oil may be simply solubilized in the hydrocarbon core of the micelles. Addition of more oil leads to

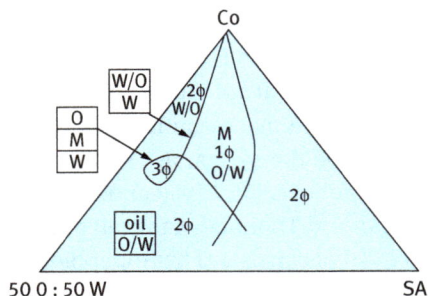

Fig. 9.4. Schematic representation of the pseudo-ternary phase diagram of oil/water/surfactant/cosurfactant.

fundamental changes of the phase diagram as is illustrated in Fig. 9.4 whereby 50 : 50 of W : O are used. To simplify the phase diagram, the 50W/50O are presented on one corner of the phase diagram.

Near the cosurfactant (co) corner the changes are small compared to the three phase diagram (Fig. 9.3). The O/W microemulsion near the water-surfactant (sa) axis is not in equilibrium with the lamellar phase, but with a non-colloidal oil + cosurfactant phase. If co is added to such a two-phase equilibrium at fairly high surfactant concentration all oil is taken up and a one-phase microemulsion appears. Addition of co at low sa concentration may lead to separation of an excess aqueous phase before all oil is taken up in the microemulsion. A three phase system is formed, containing a microemulsion that cannot be clearly identified as W/O or W/O and that is presumably similar to the lamellar phase swollen with oil or to a more irregular intertwining of aqueous and oily regions (bicontinuous or middle phase microemulsion).

The interfacial tensions between the three phases are very low ($0.1 - 10^{-4}$ mNm^{-1}). Further addition of co to the three phase system makes the oil phase disappear and leaves a W/O microemulsion in equilibrium with a dilute aqueous sa solution.

In the large one-phase region, continuous transitions from O/W to middle phase to W/O microemulsions are found.

Microemulsions can also be illustrated by considering the phase diagrams of non-ionic surfactants containing poly(ethylene oxide) (PEO) head groups, as discussed by Shinoda and Friberg [4]. Such surfactants do not generally need a cosurfactant for microemulsion formation. A schematic representation of the phase behavior of nonionic surfactants is given in Fig. 9.5

At low temperatures, the ethoxylated surfactant is soluble in water and at a given concentration is capable of solubilizing a given amount of oil. The oil solubilization increases rapidly with increasing temperature near the cloud point of the surfactant – this is illustrated in Fig. 9.5a which shows the solubilization and cloud point curves of the surfactant. Between these two curves, an isotropic region of O/W solubilized system exists. At any given temperature, any increase in the oil weight fraction above the solubilization limit results in oil separation (oil solubilized + oil). At any given

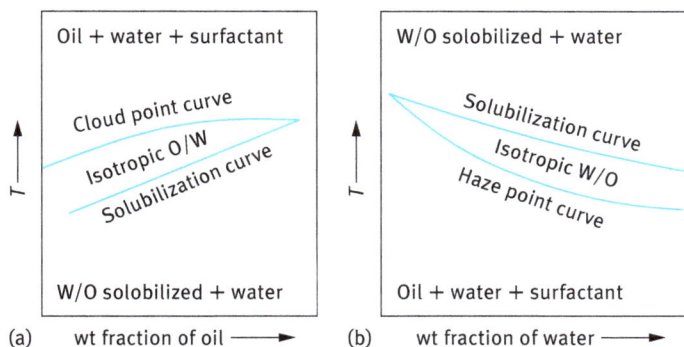

Fig. 9.5. Schematic representation of the phase behavior of nonionic surfactants: (a) oil solubilized in a nonionic surfactant solution; (b) water solubilized in an oil solution of a nonionic surfactant.

surfactant concentration, any increase in temperature above the cloud point results in separation into oil, water and surfactant.

If one starts from the oil phase with dissolved surfactant and adds water, solubilization of the latter takes place and solubilization increases with reduction of temperature near the haze point. Between the solubilization and haze point curves, an isotropic region of W/O solubilized system exists. At any given temperature, any increase in water weight fraction above the solubilization limit results in water separation (W/O solubilized + water). At any given surfactant concentration, any decrease in temperature below the haze point results in separation to water, oil and surfactant.

With nonionic surfactants, both types of microemulsions can be formed depending on the conditions. With such systems, temperature is the most crucial factor since the solubility of surfactant in water or oil depends on temperature. Microemulsion prepared using nonionic surfactants have a limited temperature range.

9.4 Thermodynamic theory of microemulsion formation

The spontaneous formation of the microemulsion with decreasing free energy can only be expected if the interfacial tension is so low that the remaining free energy of the interface is over compensated for by the entropy of dispersion of the droplets in the medium [5, 6]. This above concept forms the basis of the thermodynamic theory proposed by Ruckenstein and Chi and Overbeek [5, 6]. The ultralow interfacial tension is produced in most cases by the use of two surfactants of different nature. Single surfactants do lower the interfacial tension γ, but in most cases the critical micelle concentration (cmc) is reached before γ is close to zero. Addition of a second surfactant of a completely different nature (i.e. predominantly oil soluble such as an alcohol) then lowers γ further and very small, even transiently negative values may be reached

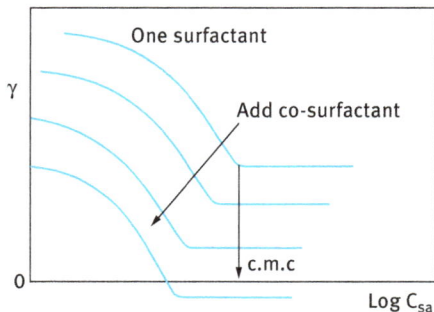

Fig. 9.6. $\gamma - \log C_{sa}$ curves for Surfactant + Cosurfactant.

[7]. This is illustrated in Fig. 9.6 which shows the effect of addition of the cosurfactant on the $\gamma - \log c_{sa}$ curve. It can be seen that addition of cosurfactant shifts the whole curve to low γ values and the cmc is shifted to lower values.

The reason for the lowering of γ when using two surfactant molecules can be understood from consideration of the Gibbs adsorption equation for multicomponent systems [7]. For a multicomponent system i, each with an adsorption Γ_i (moles m^{-2}, referred to as the surface excess), the reduction in γ, i.e. dγ, is given by the following expression:

$$d\gamma = -\sum \Gamma_i d\mu_i = -\sum \Gamma_i RT \, d\ln C_i, \tag{9.2}$$

where μ_i is the chemical potential of component i, R is the gas constant, T is the absolute temperature and C_i is the concentration (mol dm^{-3}) of each surfactant component.

For two components, sa (surfactant) and co (cosurfactant), equation (9.2) becomes

$$d\gamma = -\Gamma_{sa} RT \, d\ln C_{sa} - \Gamma_{co} RT \, d\ln C_{co}. \tag{9.3}$$

Integration of equation (9.3) gives

$$\gamma = \gamma_0 - \int_0^{C_{sa}} \Gamma_{sa} RT \, d\ln C_{sa} - \int_0^{C_{co}} \Gamma_{co} RT \, d\ln C_{co}, \tag{9.4}$$

which clearly shows that γ_0 is lowered by two terms, both from surfactant and cosurfactant.

The two surfactant molecules should adsorb simultaneously and they should not interact with each other, otherwise they lower their respective activities. Thus, the surfactant and cosurfactant molecules should vary in nature, one predominantly water soluble (such as an anionic surfactant) and predominantly oil soluble (such as a medium chain alcohol). In some cases a single surfactant may be sufficient for lowering γ far enough for microemulsion formation to become possible, e.g. Aerosol OT (sodium diethyl hexyl sulfosuccinate) and many nonionic surfactants.

9.5 Characterization of microemulsions using scattering techniques

Scattering techniques provide the most obvious methods for obtaining information on the size, shape and structure of microemulsions. The scattering of radiation, e.g. light, neutrons, X-ray, etc. by particles have been successfully applied for the investigation of many systems such as polymer solutions, micelles and colloidal particles. In all these methods, measurements can be made at sufficiently low concentration to avoid complications arising from particle-particle interactions. The results obtained are extrapolated to infinite dilution to obtain the desirable property such as the molecular weight and radius of gyration of a polymer coil, the size and shape of micelles, etc. Unfortunately the above dilution method cannot be applied for microemulsions, which depend on a specific composition of oil, water and surfactants. The microemulsions cannot be diluted by the continuous phase since this dilution results in breakdown of the microemulsion. Thus, when applying the scattering techniques to microemulsions measurements have to be made at finite concentrations and the results obtained have to be analyzed using theoretical treatments to take into account the droplet-droplet interactions.

Below, two scattering methods will be discussed: time average (static) light scattering and dynamic (quasi-elastic) light scattering referred to as photon correlation spectroscopy.

9.5.1 Time average (static) light scattering

The intensity of scattered light $I(Q)$ is measured as a function of scattering vector Q,

$$Q = \left(\frac{4\pi n}{\lambda}\right) \sin\left(\frac{\theta}{2}\right), \tag{9.5}$$

where n is the refractive index of the medium, λ is the wave length of light and θ is the angle at which the scattered light is measured.

For a fairly dilute system, $I(Q)$ is proportional to the number of particles N, the square of the individual scattering units V_p and some property of the system (material constant) such as its refractive index,

$$I(Q) = [(\text{Material const.})\,(\text{Instrument const.})]\,N\,V_p^2. \tag{9.6}$$

The instrument constant depends on the geometry of the apparatus (the light path length and the scattering cell constant).

For more concentrated systems, $I(Q)$ also depends on the interference effects arising from particle-particle interaction,

$$I(Q) = [(\text{Instrument const.})\,(\text{Material const.})]\,N\,V_p^2\,P(Q)\,S(Q), \tag{9.7}$$

where P(Q) is the particle form factor which allows the scattering from a single particle of known size and shape to be predicted as a function of Q. For a spherical particle of radius R,

$$P(Q) = \left[\frac{(3 \sin QR - QR \cos QR)}{(QR)^3} \right]^2 , \tag{9.8}$$

S(Q) is the so-called "structure factor" which takes into account the particle-particle interaction. S(Q) is related to the radial distribution function g(r) (which gives the number of particles in shells surrounding a central particle) [8],

$$S(Q) = 1 - \frac{4\pi N}{Q} \int_0^\infty [g(r) - 1] \, r \sin QR \, dr . \tag{9.9}$$

For a hard-sphere dispersion with radius R_{HS} (which is equal to R + t, where t is the thickness of the adsorbed layer),

$$S(Q) = \frac{1}{[1 - NC, (2Q \, R_{HS})]} , \tag{9.10}$$

where C is a constant.

One usually measures I(Q) at various scattering angles θ and then plots the intensity at some chosen angle (usually 90°), i_{90} as a function of the volume fraction ϕ of the dispersion. Alternatively, the results may be expressed in terms of the Rayleigh ratio R_{90},

$$R_{90} = \left(\frac{i_{90}}{I_0} \right) r_s^2 , \tag{9.11}$$

I_0 is the intensity of the incident beam and r_s is the distance from the detector.

$$R_{90} = K_0 M \, C \, P(90) \, S(90) , \tag{9.12}$$

K_0 is an optical constant (related to the refractive index difference between the particles and the medium). M is the molecular mass of scattering units with weight fraction C.

For small particles (as is the case with microemulsions) $P(90) \sim 1$ and

$$M = \frac{4}{3} \pi R_c^3 N_A , \tag{9.13}$$

where N_A is the Avogadro's constant,

$$C = \phi_c \rho_c , \tag{9.14}$$

where ϕ_c is the volume fraction of the particle core and ρ_c is its density.

Equation (9.14) can be written in the simple form

$$R_{90} = K_1 \phi_c R_c^3 S(90) , \tag{9.15}$$

where $K_1 = K_0 (4/3) N_A \rho_c^2$.

Equation (9.15) shows that to calculate R_c from R_{90} one needs to know S(90). The latter can be calculated using equations (9.9) and (9.10).

The above calculations were obtained using a W/O microemulsion of water/ xylene/sodium dodecyl benzene sulfonate (NaDBS)/hexanol [9]. The microemulsion region was established using the quaternary phase diagram. W/O microemulsions were produced at various water volume fractions using increasing amounts of NaDBS: 5, 10.9, 15 and 20 %.

The results for the variation of R_{90} with the volume fraction of the water core droplets at various NaDBS concentrations are shown in Fig. 9.7. With the exception of the 5 % NaDBS results, all the others showed an initial increase in R_{90} with increasing ϕ, reaching a maximum at a given ϕ, after which R_{90} decreases with a further increase in ϕ.

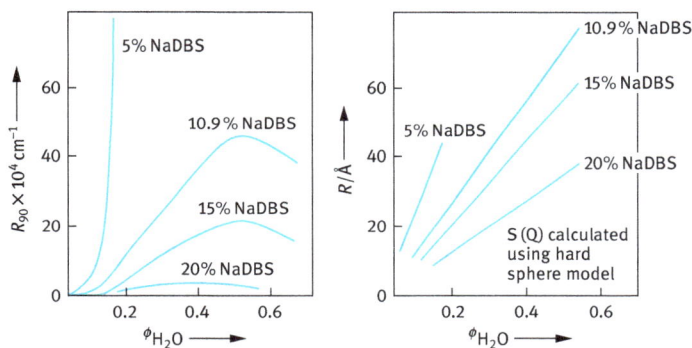

Fig. 9.7. Variation of R_{90} and R with the water volume fraction for a W/O microemulsion based on xylene-ware-NaDBS-hexanol.

The above results were used to calculate R as a function of ϕ using the hard-sphere model discussed above (equation (9.15)). This is also shown in Fig. 9.7. It can be seen that with increasing ϕ, at constant surfactant concentration, R increases (the ratio of surfactant to water decreases with increasing ϕ). At any volume fraction of water, an increase in surfactant concentration results in a decrease in the microemulsion droplet size (the ratio of surfactant to water increases).

9.5.1.1 Calculation of droplet size from interfacial area

If one assumes that all surfactant and cosurfactant molecules are adsorbed at the interface, it is possible to calculate the total interfacial area of the microemulsion from a knowledge of the area occupied by surfactant and cosurfactant molecules:

Total interfacial area = Total number of surfactant molecule$_s$ × area per surfactant molecule A_s + total number of cosurfactant molecules × area per cosurfactant molecule A_{co}.

The total interfacial area A per Kg of microemulsion is given by the expression,

$$A = \frac{(n_s\, N_A\, A_s + n_{co}\, N_A\, A_{co})}{\phi},$$ (9.16)

n_s and n_{co} are the number of moles of surfactant and cosurfactant.

A is related to the droplet radius R (assuming all the droplets are of the same size) by

$$A = \frac{3}{R\rho}.$$ (9.17)

Using reasonable values for A_s and A_{co} (30 A^2 for NaDBS and 20 A^2) for hexanol) R was calculated and the results were compared with those obtained using light scattering results. Good agreement was obtained between the two sets of results.

9.5.2 Dynamic light scattering (photon correlation spectroscopy, PCS)

In this technique one measures the intensity fluctuation of scattered light by the droplets as they undergo Brownian motion [10]. When a light beam passes through a colloidal dispersion, an oscillating dipole movement is induced in the particles, thereby radiating the light. Due to the random position of the particles, the intensity of scattered light, at any instant, appears as random diffraction ("speckle" pattern). As the particles undergo Brownian motion, the random configuration of the pattern will fluctuate, such that the time taken for an intensity maximum to become a minimum (the coherence time), corresponds approximately to the time required for a particle to move one wavelength λ. Using a photomultiplier of active area about the diffraction maximum (i.e. one coherent area) this intensity fluctuation can be measured. The analogue output is digitized (using a digital correlator) that measures the photocount (or intensity) correlation function of scattered light.

The photocount correlation function $g^{(2)}(\tau)$ is given by

$$g^{(2)} = B[1 + \gamma^2 g^{(1)}(\tau)]^2$$ (9.18)

where τ is the correlation delay time. B is the Background value to which $g^{(2)}(\tau)$ decays at long delay times. $g^{(2)}(\tau)$ is the normalized correlation function of the scattered electric field and γ is a constant (~ 1).

The correlator compares $g^{(2)}(\tau)$ for many values of τ. For monodispersed noninteracting particles,

$$g^{(1)}(\tau) = \exp(-\Gamma\gamma)$$ (9.19)

Γ is the decay rate or inverse coherence time that is related to the translational diffusion coefficient D,

$$\Gamma = D K^2 \tag{9.20}$$

where K is the scattering vector,

$$K = \left(\frac{4\pi n}{\lambda_o}\right) \sin\left(\frac{\theta}{2}\right). \tag{9.21}$$

The particle radius R can be calculated from D using the Stokes–Einstein equation,

$$D = \frac{kT}{6\pi \eta_o R}, \tag{9.22}$$

where η_o is the viscosity of the medium.

The above analysis only applies for very dilute dispersions. With microemulsions which are concentrated dispersions, corrections are needed to take into account the interdroplet interaction. This is reflected in plots of $\ln g^{(1)}(\tau)$ versus τ which become nonlinear, implying that the observed correlation functions are not single exponentials.

As with time average light scattering, one needs to introduce a structure factor in calculating the average diffusion coefficient. For comparative purposes, one can calculate the collective diffusion coefficient D which can be related to its value at infinite dilution D_o by [11],

$$D = D_o(1 + \alpha, \phi) \tag{9.23}$$

where α is a constant that is equal to 1.5 for hard spheres with repulsive interaction.

9.6 Characterization of microemulsions using conductivity

Conductivity measurements may provide valuable information on the structural behavior of microemulsions. In the early applications of conductivity measurements, the technique was used to determine the nature of the continuous phase. O/W microemulsions should give fairly high conductivity (which is determined by that of the continuous aqueous phase) whereas W/O microemulsions should give fairly low conductivity (which is determined by that of the continuous oil phase).

As an illustration, Fig. 9.8 shows the change in electrical resistance (reciprocal of conductivity) with the ratio of water to oil (V_w/V_o) for a microemulsion system prepared using the inversion method [2]. Fig. 9.8 indicates the change in optical clarity and birifringence with the ratio of water to oil.

At low V_w/V_o, a clear W/O microemulsion is produced with a high resistance (oil continuous). As V_w/V_o increases, the resistance decreases, and in the turbid region, hexanol and lamellar micelles are produced. Above a critical ratio, inversion occurs and the resistance decreases producing O/W microemulsion.

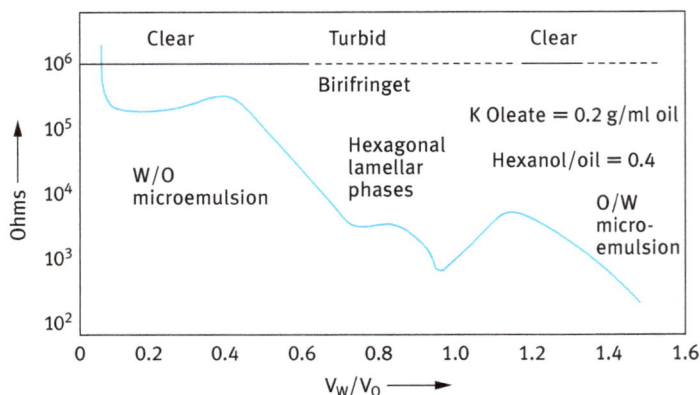

Fig. 9.8. Electrical resistance versus V_w/V_o.

Conductivity measurements were also used to study the structure of the microemulsion, which is influenced by the nature of the cosurfactant. A systematic study of the effect of cosurfactant chain length on the conductive behavior of W/O microemulsions was carried out by Clausse and his coworkers [12, 13]. The cosurfactant chain length was gradually increased from C_2 (Ethanol) to C_7 (Heptanol). The results for the variation of κ with ϕ_w are shown in Fig. 9.9.

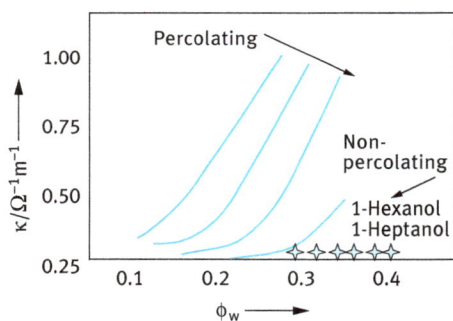

Fig. 9.9. Variation of conductivity with water volume fraction for various cosurfactants.

With the short chain alcohols ($C < 5$), the conductivity shows a rapid increase above a critical ϕ value. With longer chain alcohols, namely hexanol and heptanol, the conductivity remains very low up to a high water volume fraction. With the short chain alcohols, the system shows percolation above a critical water volume fraction. Under these conditions the microemulsion is "bicontinuous". With the longer chain alcohols, the system is non-percolating and one can define definite water cores. This is sometimes referred to as a "true" microemulsion.

9.7 NMR measurements

Lindman and coworkers [14–16] demonstrated that the organization and structure of microemulsions can be elucidated from self-diffusion measurements of all the components (using pulse gradient or spin echo NMR techniques). Within a micelle, the molecular motion of the hydrocarbon tails (translational, reorientation and chain flexibility) is almost as rapid as in a liquid hydrocarbon. In a reverse micelle, water molecules and counterions are also highly mobile. For many surfactant-water systems, there is a distinct spatial separation between hydrophobic and hydrophilic domains. The passage of species between different regions is an improbable event and this occurs very slowly.

Thus, self-diffusion, if studied over macroscopic distances, should reveal whether the process is rapid or slow depending on the geometrical properties of the inner structure. For example, a phase that is water continuous and oil discontinuous should exhibit rapid diffusion of hydrophilic components, while the hydrophobic components should diffuse slowly. An oil continuous but water discontinuous system should exhibit rapid diffusion of the hydrophobic components. One would expect that a bicontinuous structure should give rapid diffusion of all components.

Using the above principle, Lindman and coworkers [14–16] measured the self-diffusion coefficients of all components consisting of various components, with particular emphasis to the role of the cosurfactant. For microemulsions consisting of water, hydrocarbon, an anionic surfactant and a short chain alcohol (C_4 and C_5), the self-diffusion coefficient of water, hydrocarbon and cosurfactant was quite high, of the order of 10^{-9} m^2s^{-1}, i.e. two orders of magnitude higher than the value expected for a discontinuous medium (10^{-11} m^2s^{-1}). This high diffusion coefficient was attributed to three main effects: bicontinuous solutions, easily deformable and flexible interfaces and absence of any large aggregates. With microemulsions based on long chain alcohols (e.g. decanol), the self-diffusion coefficient for water was low, indicating the presence of definite (closed) water droplets surrounded by surfactant anions in the hydrocarbon medium. Thus, NMR measurements could clearly distinguish between the two types of microemulsion systems.

9.8 Formulation of microemulsions

The formulation of microemulsions or micellar solutions, like that of conventional macroemulsions is still an art. In spite of the exact theories that explain the formation of microemulsions and their thermodynamic stability, the science of microemulsion formulation has not advanced to the point where one can predict with accuracy what happens when the various components are mixed. The very much higher ratio of emulsifier to disperse phase which differentiates microemulsions from macroemulsions appears at first sight for the application of various techniques for formulation to be less critical. However, in the final stages of the formulation one immediately realizes that the requirements are very critical due to the greater number of parameters involved.

The mechanics of forming microemulsions differ from those used in making macroemulsions. The most important difference lies in the fact that putting more work into a macroemulsion or increasing emulsifier usually improves its stability. This is not so for microemulsions. Formation of a microemulsion depends on specific interactions of the molecules of oil, water and emulsifiers. These interactions are not exactly known. If such specific interactions are not realized, no amount of work nor excess emulsifier can produce the microemulsion. If the chemistry is right, microemulsification occurs spontaneously.

One should remember that for microemulsions the ratio of emulsifier to oil is much higher than that used for macroemulsions. This emulsifier used is at least 10 % based on the oil and in most cases it can be as high as 20–30 %. The W/O systems are made by blending the oil and emulsifier with some heating if necessary. Water is added to the oil-emulsifier blend to produce the microemulsion droplets and the resulting system should appear transparent or translucent. If the maximum amount of water that can be microemulsified is not high enough for the particular application, one should try other emulsifiers to reach the required composition. The most convenient way of producing O/W microemulsion is to blend the oil and emulsifier and the pour the mixture into water with mild stirring. In the case of waxes, both oil/emulsifier blend and the water must be at higher temperature (above the melting point of the wax). If the melting point of the wax is above the boiling temperature of water, the process can be carried out at high pressure. Another technique to mix the ingredients is to make a crude macroemulsion of the oil and one of the emulsifiers. By using low volumes of water, a gel is formed and the system can then be titrated with the co-emulsifier till a transparent system is produced. The above system may be further diluted with water to produce a translucent microemulsion.

Three different emulsifier selection methods can be applied for formulation of microemulsions:
1. the hydrophilic-lipophilic-balance (HLB) system;
2. the phase inversion temperature (PIT) method;
3. partitioning of cosurfactant between the oil and water phases.

The first two methods are essentially the same as those used for selection of emulsifiers for macroemulsions and these were described in Chapter 6. However, with microemulsions one should try to match the chemical type of the emulsifier with that of the oil.

According to the thermodynamic theory of microemulsion formation, the total interfacial tension of the mixed film of surfactant and cosurfactant must approach zero.

The total interfacial tension is given by the following equation:

$$\gamma_i = (\gamma_{O/W})_a - \pi \qquad (9.24)$$

where $(\gamma_{O/W})_a$ is the interfacial tension of the oil in the presence of alcohol cosurfactant and π is the surface pressure. $(\gamma_{O/W})_a$ seems to reach a value of 15 mNm^{-1} irrespective of the original value of $\gamma_{O/W}$. It seems that the cosurfactant which is predominantly oil soluble distributes itself between the oil and the interface and this causes a change in the composition of the oil which now is reduced to 15 mNm^{-1}.

Measurement of the partition of the cosurfactant between the oil and the interface is not easy. A simple procedure to select the most efficient cosurfactant is to measure the oil/water interfacial tension $\gamma O/W$ as a function of cosurfactant concentration. The lower the percentage of cosurfactant required to lower $\gamma_{O/W}$ to 15 mNm^{-1} the better the candidate.

References

[1]   T. P. Hoar and J. H. Schulman, *Nature* (London) **152**, 102 (1943).

[2]   L. M. Prince, *Microemulsion Theory and Practice*, Academic Press, New York, 1977.

[3]   I. Danielsson, and B. Lindman, *Colloids and Surfaces*, **3**, 391 (1983).

[4]   K. Shinoda and S. Friberg, *Adv. Colloid Interface Sci.*, **4**, 281 (1975).

[5]   E. Ruckenstein and J. C. Chi, *J. Chem. Soc. Faraday Trans. II*, **71**, 1690 (1975).

[6]   J. Th. G. Overbeek, *Faraday Disc. Chem. Soc.*, **65**, 7 (1978).

[7]   J. T. G. Overbeek, P. L. de Bruyn and F. Verhoeckx, in: *Surfactants*, Th. F. Tadros (ed.), pp. 111–132, Academic Press, London, 1984.

[8]   R. C. Baker, A. T. Florence, R. H. Ottewill and Th. F. Tadros, *J. Colloid Interface Sci.*, **100**, 332 (1984).

[9]   N. W. Ashcroft and J. Lekner, *Phys. Rev.* **45**, 33 (1966).

[10]   P. N. Pusey, in: *Industrial Polymers: Characterisation by Molecular Weights*, J. H. S. Green and R. Dietz (eds.), Transcripta Books, London, 1973.

[11]   A. N. Cazabat and D. Langevin, *J. Chem. Phys.*, **74**, 3148 (1981).

[12]   B. Lagourette, J. Peyerlasse, C. Boned and M. Clausse, *Nature*, **281**, 60 (1969).

[13]   M. Clausse, J. Peyerlasse, C. Boned, J. Heil, L. Nicolas-Margantine and A. Zrabda, in: *Solution Properties of Surfactants*, Vol. 3, K. L. Mittal and B. Lindman (eds.), p. 1583, Plenum Press, New York, 1984.

[14]   B. Lindman and H. Winnerstrom, in: *Topics in Current Chemistry*, F. L. Borschke (ed.), pp. 1–83, Springer-Verlag, Heidelberg, 1980.

[15]   H. Winnerstrom and B. Lindman, *Phys. Rep.*, **52**, 1 (1970).

[16]   B. Lindman, P. Stilbs and M. E. Moseley, *J. Colloid Interface Sci.*, **83**, 569 (1981).

10 Surfactants as wetting agents

10.1 Introduction

Wetting is important in many industrial processes and in many cases complete wetting is a prerequisite for applications e.g. in paint application where the paint has to wet the substrate completely in order to from a uniform paint film. In crop sprays applied to plants or weeds, it is essential that the spray solution wets the substrate completely and in many cases rapid spreading may be required. In this case, the dynamics of wetting becomes a very important factor. In personal care formulations such as creams and lotions, good wetting of the substrate (skin) is required. Also in hair sprays, droplet impaction and adhesion become important and this may have to be followed by wetting and spreading on the hair surface. In pharmaceutical applications, wetting of tablets is essential for their disintegration and dispersion.

Wetting of powders is an important prerequisite for dispersion of powders in liquids, i.e. preparation of suspensions. It is essential to wet both the external and internal surfaces of the powder aggregated and agglomerates. Suspensions are applied in many industries such as paints, dyestuffs, printing inks, agrochemicals, pharmaceuticals, paper coatings, detergents, etc.

In all the above processes one has to consider both the equilibrium and dynamic aspects of the wetting process [1]. The equilibrium aspects of wetting can be studied at a fundamental level using interfacial thermodynamics. Under equilibrium, a drop of a liquid on a substrate produces a contact angle θ, which is the angle formed between planes tangent to the surfaces of solid and liquid at the wetting perimeter. This is illustrated in Fig. 10.1 which shows the profile of a liquid drop on a flat solid substrate. An equilibrium between vapor, liquid and solid is established with a contact angle θ (that is lower than 90°).

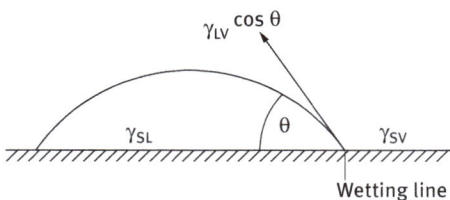

Fig. 10.1. Schematic representation of the contact angle and wetting lane.

The wetting perimeter is frequently referred to as the three-phase line (Solid/Liquid/Vapor); the most common name is the wetting line. Most equilibrium wetting studies center around measurements of the contact angle; the smaller the angle the better the liquid is said to wet the solid. Typical examples are given in Table 10.1 for water with a surface tension of $72\,\mathrm{mNm^{-1}}$ on various substrates.

Table 10.1. Typical contact angle values for a water drop on various substrates.

Substrate Contact	angle θ/°
PTFE (Teflon)	112
Paraffin Wax	110
Polyethylene	103
Human Skin	75–90
Glass	0

The above values can be roughly used as a measure of wetting of the substrate by water (glass being completely wetted and PTFE very difficult to wet).

The spreading of liquids on substrates is also an important industrial phenomenon, for example as in the case with crop sprays which need to spread spontaneously on leaf surfaces in order to maximize the biological effect. A useful concept introduced is the spreading coefficient which will be described below

10.2 The concept of contact angle

Wetting is a fundamental interfacial phenomenon in which one fluid phase is displaced completely or partially by another fluid phase from the surface of a solid or a liquid. The most useful parameter than may describe wetting is the contact angle of a liquid on the substrate. When a drop of a liquid is placed onto a solid, the liquid either spreads to form a thin (uniform) film or remains as a discrete drop. This is schematically illustrated in Fig. 10.2.

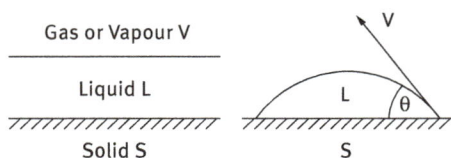

Gas or Vapour V

Liquid L

Solid S

Complete wetting

Liquid remains as a descrete drop incomplete wetting

Fig. 10.2. Illustration of complete and partial wetting.

The contact angle θ is the angle formed between planes tangent to the surfaces of the solid and liquid at the wetting perimeter. As mentioned before, the wetting perimeter is referred to as the three-phase line (solid/liquid/vapor) or simply the wetting line. The utility of contact angle measurements depends on equilibrium thermodynamic arguments (static measurements). In practical systems such as in spray applications, one has to displace one fluid (air) with another (liquid) as quickly and as efficiently as possible. Dynamic contact angle measurements (associated with a moving wetting line) are more relevant in many practical applications. Even under static conditions, contact angle measurements are far from being simple since they are mostly accompanied by hysteresis. The value of θ depends on the history of the system and whether the liquid is tending to advance across or recedes from the solid surface. The limiting angles achieved just prior to movement of the wetting line (or just after movement ceases) are known as the advancing and receding contact angles, θ_A and θ_R, respectively. For a given system $\theta_A > \theta_R$ and θ can usually take any value between these two limits without discernible movement of the wetting line.

The liquid drop takes the shape that minimizes the free energy of the system. Consider a simple system of a liquid drop (L) on a solid surface (S) in equilibrium with the vapor of the liquid (V) as was illustrated in Fig. 10.1. The sum $(\gamma_{SV}A_{SV} + \gamma_{SL}A_{SL} + \gamma_{LV}A_{LV})$ should be a minimum at equilibrium and this leads to the Young's equation [2],

$$\gamma_{SV} = \gamma_{SL} + \gamma_{LV}\cos\theta. \tag{10.1}$$

In the above equation, θ is the equilibrium contact angle. The angle which a drop assumes on a solid surface is the result of the balance between the cohesion force in the liquid and the adhesion force between the liquid and solid, i.e.

$$\gamma_{LV}\cos\theta = \gamma_{SV} - \gamma_{SL} \tag{10.2}$$

or

$$\cos\theta = \frac{\gamma_{SV} - \gamma_{SL}}{\gamma_{LV}}. \tag{10.3}$$

If there is no interaction between solid and liquid, then

$$\gamma_{SL} = \gamma_{SV} + \gamma_{LV} \tag{10.4}$$

i.e. $\theta = 180°$ $(\cos\theta = -1)$.

If there is strong interaction between solid and liquid (maximum wetting), the latter spreads until Young's equation is satisfied $(\theta = 0)$ and

$$\gamma_{LV} = \gamma_{SV} - \gamma_{SL}. \tag{10.5}$$

The liquid spreads spontaneously on the solid surface.

When the surface of the solid is in equilibrium with the liquid vapor, then one must consider the spreading pressure, π_e. As a result of the adsorption of the vapor on the solid surface, its surface tension γ_s is reduced by π_e, i.e.,

$$\gamma_{SV} = \gamma_s - \pi_e \tag{10.6}$$

and Young's equation can be written as,

$$\gamma_{LV}\cos\theta = \gamma_s - \gamma_{SL} - \pi_e. \tag{10.7}$$

In general, Young's equation provides a precise thermodynamic definition of the contact angle. However, it suffers from a lack of direct experimental verification since both γ_{SV} and γ_{SL} cannot be directly measured. An important criterion for application of Young's equation is to have a common tangent at the wetting line between the two interfaces.

10.3 Adhesion tension

There is no direct way by which γ_{SV} or γ_{SL} can be measured. The difference between γ_{SV} and γ_{SL} can be obtained from contact angle measurements. This difference is referred to as the "Wetting Tension" or "Adhesion Tension",

$$\text{Adhesion Tension} = \gamma_{SV} - \gamma_{SL} = \gamma_{LV} \cos\theta. \tag{10.8}$$

Thus, the adhesion tension depends on the measurable quantities γ_{LV} and θ. As long as θ is $< 90°$, the adhesion tension is positive.

10.4 Work of adhesion W_a

Consider a liquid drop with surface tension γ_{LV} and a solid surface with surface tension γ_{SV}. When the liquid drop adheres to the solid surface it forms a surface tension $\gamma\gamma_{SL}$. The work of adhesion [3, 4] is simply the difference between the surface tensions of the liquid/vapor and solid/vapor and that of the solid/liquid,

$$W_a = \gamma_{SV} + \gamma_{LV} - \gamma_{SL}. \tag{10.9}$$

Using Young's equation,

$$W_a = \gamma_{LV} (\cos\theta + 1). \tag{10.10}$$

10.5 Work of cohesion

The work of cohesion W_c is the work of adhesion when the two phases are the same. Consider a liquid cylinder with unit cross-sectional area. When this liquid is subdivided into two cylinders, two new surfaces are formed. The two new areas will have a surface tension of $2\gamma_{LV}$ and the work of cohesion is simply,

$$W_c = 2\gamma_{LV}. \tag{10.11}$$

Thus, the work of cohesion is simply equal to twice the liquid surface tension. An important conclusion may be drawn if one considers the work of adhesion given by equation (10.10) and the work of cohesion given by equation (10.11): when $W_c = W_a$, $\theta = 0°$.

This is the condition for complete wetting. When $W_c = 2W_a$, $\theta = 90°$ and the liquid forms a discrete drop on the substrate surface. Thus, the competition between the cohesion of the liquid to itself and its adhesion to a solid gives an angle of contact that is constant and specific to a given system at equilibrium. This shows the importance of Young's equation in defining wetting.

10.6 The spreading coefficient S

Harkins [5, 6] defined the initial spreading coefficient as the work required to destroy unit area of solid/liquid (SL) and liquid/vapor (LV) and leave unit area of bare solid (SV).

$$S = \gamma_{SV} - (\gamma_{SL} + \gamma_{LV}) \tag{10.12}$$

$$S = \gamma_{LV} (\cos \theta + 1) . \tag{10.13}$$

If S is positive, the liquid will spread until it completely wets the solid so that $\theta = 0°$. If S is negative ($\theta > 0°$) only partial wetting occurs. Alternatively, one can use the equilibrium or final spreading coefficient.

10.7 Contact angle hysteresis

For a liquid spreading on a uniform, non-deformable solid (idealized case), there is only one contact angle (the equilibrium value). With real surfaces (practical systems) a number of stable angles can be measured. Two relatively reproducible angles can be measured: largest, advancing angle θ_A and smallest, receding angle θ_R. θ_A is measured by advancing the periphery of the drop over the surface (e.g. by adding more liquid to the drop). θ_R is measured by pulling the liquid back (e.g. by removing some liquid from the drop). The difference between θ_A and θ_{3R} is termed "contact angle hysteresis".

Several factors can be considered to account for contact angle hysteresis such as penetration of wetting liquid into pores during advancing contact angle measurements and surface roughness.

Wenzel [7] considered the true area of a rough surface A (which takes into account all the surface topography, peaks and valleys) and the projected area A' (the macroscopic or apparent area). A roughness factor r can be defined as

$$r = \frac{A}{A'} , \tag{10.14}$$

$r > 1$, the higher the value of r the higher the roughness of the surface.

The measured contact angle θ (the macroscopic angle) can be related to the intrinsic contact angle θ_o through r,

$$\cos \theta = r \cos \theta_o . \tag{10.15}$$

Using Young's equation,

$$\cos\theta = r\left(\frac{\gamma_{SV} - \gamma_{SL}}{\gamma_{LV}}\right).\tag{10.16}$$

If $\cos\theta$ is negative on a smooth surface ($\theta > 90°$), it becomes more negative on a rough surface; θ becomes larger and surface roughness reduces wetting. If $\cos\theta$ is positive on a smooth surface ($\theta < 90°$), it becomes more positive on a rough surface; θ is smaller and surface roughness enhances wetting.

Another factor that can cause hysteresis is surface heterogeneity. Most real surfaces are heterogeneous consisting of patches (islands) that vary in their degrees of hydrophilicity/hydrophobicity. As the drop advances on such a heterogeneous surface, the edge of the drop tends to stop at the boundary of the island. The advancing angle will be associated with the intrinsic angle of the high contact angle region (the more hydrophobic patches or islands). The receding angle will be associated with the low contact angle region, i.e. the more hydrophilic patches or islands.

If the heterogeneities are small compared with the dimensions of the liquid drop, one can define a composite contact angle. Cassie [8, 9] considered the maximum and minimum values of the contact angles and used the following simple expression,

$$\cos\theta = Q_1 \cos\theta_1 + Q_2 \cos\theta_2,\tag{10.17}$$

Q_1 is the fraction of the surface having contact angle θ_1 and Q_2 is the fraction of the surface having contact angle θ_2. θ_1 and θ_2 are the maximum and minimum contact angles respectively.

10.8 Critical surface tension of wetting

A systematic way of characterizing "wettability" of a surface was introduced by Fox and Zisman [10]. The contact angle exhibited by a liquid on a low energy surface is largely dependent on the surface tension of the liquid γ_{LV}. For a given substrate and a series of related liquids (such as n-alkanes, siloxanes or dialkyl ethers) $\cos\theta$ is a linear function of the liquid surface tension γ_{LV}. This is illustrated in Fig. 10.3 for a number

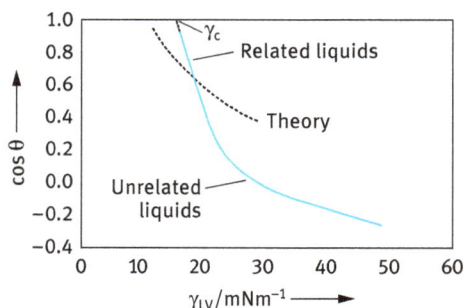

Fig. 10.3. Variation of $\cos\theta$ with γ_{LV} for related and unrelated liquids on PTFE.

of related liquids on polytetrafluoroethylene (PTFE). The figure also shows the results for unrelated liquids with widely ranging surface tensions; the line broadens into a band which tends to be curved for high surface tension polar liquids.

The surface tension at the point where the line cuts the $\cos\theta = 1$ axis is known as the critical surface tension of wetting. γ_c is the surface tension of a liquid that would just spread on the substrate to give complete wetting

The above linear relationship can be represented by the following empirical equation:

$$\cos\theta = 1 + b(\gamma_{LV} - \gamma_c). \tag{10.18}$$

High energy solids such as glass and polyethylene terphthalate have high critical surface tension ($\gamma_c > 40$ mNm^{-1}). Lower energy solids such as polyethylene have lower values of γc (~ 31 mNm^{-1}). The same applies to hydrocarbon surfaces such as paraffin wax. Very low energy solids such as PTFE have lower γ_c of the order of 18 mNm^{-1}. The lowest known value is ~ 6 mNm^{-1}, which is obtained using condensed monolayers of perfluorolauric acid.

10.9 Effect of surfactant adsorption

Surfactants lower the surface tension of the liquid, γ_{LV}, and they also adsorb at the solid/liquid interface lowering γ_{SL}. The adsorption of surfactants at the liquid/air interface can be easily described by the Gibbs adsorption equation [11],

$$\frac{d\gamma_{LV}}{dC} = -2.303\Gamma\,RT, \tag{10.19}$$

where C is the surfactant concentration (moles dm^{-3}) and Γ is the surface excess (amount of adsorption in moles m^{-2}).

Γ can be obtained from surface tension measurements using solutions with various molar concentrations (C). From a plot of γ_{LV} versus log C one can obtain Γ from the slope of the linear portion of the curve just below the critical micelle concentration (cmc).

The adsorption of surfactant at the solid /liquid interface also lowers $_{SL}$. From Young's equation,

$$\cos\theta = \frac{\gamma_{SV} - \gamma_{SL}}{\gamma_{LV}}. \tag{10.20}$$

Surfactants reduce θ if either γ_{SL} or γ_{LV} or both are reduced (when γ_{SV} remains constant). Smolders [12] obtained an equation for the change of contact angle with surfactant concentration by differentiating Young's equation with respect to ln C at constant temperature,

$$\frac{d(\gamma_{LV}\cos\theta)}{d\ln C} = \frac{d\gamma_{SV}}{d\ln C} - \frac{d\gamma_{SL}}{d\ln C}. \tag{10.21}$$

Using the Gibbs equation,

$$\sin \theta \left(\frac{d\theta}{dl\, C} \right) = RT(\Gamma_{SV} - \Gamma_{SL} - \Gamma_{LV} \cos \theta). \tag{10.22}$$

Since $\gamma_{LV} \sin \theta$ is always positive, then $(d\theta/d\ln C)$ will always have the same sign as the right-hand side of equation (10.22) and three cases may be distinguished:

1. $(d\theta/d\ln C) < 0$; $\Gamma_{SV} < \Gamma_{SL} + \Gamma_{LV} \cos \theta$. Addition of surfactant improves wetting.
2. $(d\theta/d\ln C) = 0$; $\Gamma_{SV} = \Gamma_{SL} + \Gamma_{LV} \cos \theta$ (no effect)
3. $(d\theta/d\ln C) > 0$; $\Gamma_{SV} > \Gamma_{SL} + \Gamma_{LV} \cos \theta$. Addition of surfactant causes dewetting.

10.10 Measurement of contact angles

The most common method for measuring the contact angle is the sessile drop or adhering gas bubble procedure. A schematic representation of a sessile drop on a flat surface and an air bubble resting on a solid surface is given in Fig. 10.4.

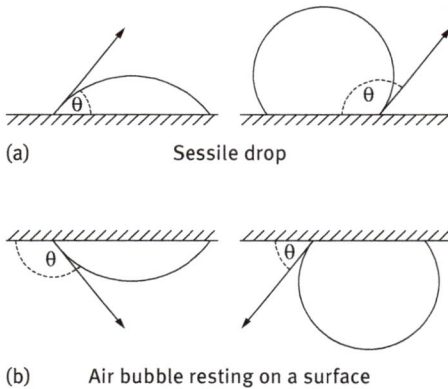

(a)   Sessile drop

(b)   Air bubble resting on a surface

Fig. 10.4. Schematic representation of the sessile drop (a) and air bubble (b) resting on a surface.

The contact angle can be measured using a telescope fitted with a goniometer eye piece. Alternatively, it can be measured by taking a photograph or using image analysis. The accuracy of measurement is $+2°$ for θ values between $10°$ and $160°$. For $\theta < 10°$ or $> 160°$, uncertainty is higher and θ can be calculated from the drop profile (applicable to drops $< 10^{-4}$ ml). This is schematically shown in Fig. 10.5.

$$\tan \left(\frac{\theta}{2} \right) = \frac{2\,h}{d} \tag{10.23}$$

$$\frac{d^3}{V} = \frac{24\sin^3\theta}{\pi(2 - 3\cos\theta + \cos^3\theta)}. \tag{10.24}$$

Care must be taken for kinetic effects and evaporation.

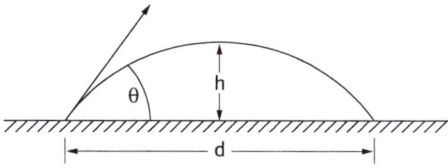

Fig. 10.5. Drop profile for calculation of contact angle.

References

[1]   T. B. Blake, in: *Surfactants*, Th. F. Tadros (ed.), Academic Press, London, 1984.
[2]   T. Young, *Phil. Trans. Royal Soc.* (London), **95**, 65 (1805).
[3]   D. H. Everett, *Pure and Appl. Chemistry.*, **52**, 1279 (1980).
[4]   R. E. Johnson, *J. Phys. Chem.*, **63**, 1655 (1959).
[5]   W. D. Harkins, *J. Phys. Chem.*, **5**, 135 (1937)
[6]   W. D. Harkins, *The Physical Chemistry of Surface Films*, Reinhold, New York, 1952.
[7]   R. N Wenzel, *Ind. Eng. Chem.*, **28**, 988 (1936).
[8]   A. B. D. Cassie and S. Dexter, *Trans. Faraday Soc.*, **40**, 546 (1944)
[9]   A. B .D. Cassie, *Disc. Faraday Soc.*, **3**, 361 (1948).
[10] H. W. Fox and W. A. Zisman, *J. Colloid Sci.*, **7**, 109, 428 (1952).
[11] J. W. Gibbs, *The Collected Work of J. Willard Gibbs*, Vol. 1, Longman-Green, New York, 1928.
[12] C. A. Smolders, *Rec. Trav. Chim.*, **80**, 650 (1960).

11 Industrial applications of surfactants

11.1 Surfactants in the home, personal care and cosmetics [1, 2]

Surfactants used in home, personal care and cosmetic formulations must be completely free of allergens, sensitizers and irritants. Conventional surfactants of the anionic, cationic, amphoteric and nonionic types are used in these formulations. Besides the synthetic surfactants that are used in preparation of many systems such as emulsions, creams, suspensions, etc., several other naturally occurring materials have been introduced and there is a trend in recent years to used such natural products more widely, in the belief that they are safer for application.

Several synthetic surfactants that are applied in home, personal care and cosmetic formulations may be listed, such as carboxylates, ether sulfates, sulfate, sulfonates, quaternary amines, betaines, sarcosinates, etc. The ethoxylated surfactants are perhaps the most widely used emulsifiers in these formulations. Being uncharged, these molecules have a low skin sensitization potential. This is due to their low binding to proteins. Unfortunately, one of the problems of nonionic surfactants is the formation of dioxane that is produced from any residual-free ethylene oxide, which even in small quantities is unacceptable due to its carcinogeneity. It is, therefore, important when using ethoxylated surfactants to ensure that the level of the free monomer is kept at very low concentration to avoid any side effects. Another drawback of ethoxylated surfactants is their degradation by oxidation or photo-oxidation processes. These problems are reduced by using sucrose esters obtained by esterification of the sugar hydroxyl groups with fatty acids such as lauric and stearic acid. In this case, the danger of dioxane contamination is absent and they are still mild to the skin, since they do not interact to any appreciable extent with proteins.

Another class of surfactants that are used in personal care and cosmetic formulations are the phosphoric acid esters. These molecules are similar to the phospholipids that constitute the natural building blocks of the stratum corneum. Glycerine esters, in particular the triglycerides, are also used in many cosmetic formulations. These surfactants are important ingredients of sebum, the natural lubricant of the skin. Being naturally occurring, they are claimed to be very safe, causing practically no medical hazard. In addition, these triglycerides can be prepared with a large variety of substituents and hence their hydrophilic-lipophilic-balance (HLB) values can be varied over a wide range.

The macromolecular surfactants possess considerable advantages for use in cosmetic ingredients. The most commonly used materials are the ABA block copolymers, with A being poly(ethylene oxide) and B poly(propylene oxide) (Pluronics). On the whole, polymeric surfactants have much lower toxicity, sensitization and irritation potentials, provided they are not contaminated with traces of the parent monomers.

Several natural surfactants are used in cosmetic formulations, such as those produced from lanolin (wool fat), phytosteroids extracted from various plants and surfactants extracted from beeswax. Unfortunately, these naturally occurring surfactants are not widely used in cosmetics due to their relatively poor physicochemical performance when compared with synthetic molecules.

Another important class of natural surfactants are the proteins, e.g. casein in milk. As with macromolecular surfactants, proteins adsorb strongly and irreversibly at the oil-water interface and hence they can stabilize emulsions effectively. However, the high molecular weight of proteins and their compact structures make them unsuitable for preparation of emulsions with small droplet sizes. For this reason, many proteins are modified by hydrolysis to produce lower molecular weight protein fragments, e.g. polypeptides, or by chemical alteration of the reactive protean side chains. Protein-sugar condensates are sometimes used in skin care formulations. In addition, these proteins impart to the skin a lubricous feel and can be used as moisturizing agent.

In recent years, there has been a great trend towards using silicone oils for many cosmetic formulations. In particular, volatile silicone oils have found application in many cosmetic products, owing to the pleasant dry sensation they impart to the skin. These volatile silicones evaporate without unpleasant cooling effects or without leaving a residue. Due to their low surface energy, silicone oils help spread the various active ingredients over the surface of hair and skin. The chemical structure of the silicone compounds used in cosmetic preparations varies according to the application. The backbones can carry various attached "functional" groups, e.g. carboxyl, amine, sulfhydryl, etc. While most silicone oils can be emulsified using conventional hydrocarbon surfactants, there has been a trend in recent years to use silicone surfactants for producing the emulsion. The surface activity of these block copolymers depends on the relative length of the hydrophobic silicone backbone and the hydrophilic (e.g. PEO) chains. The attraction of using silicone oils and silicone copolymers is their relatively small medical and environmental hazards, when compared to their hydrocarbon counterparts.

Several examples of personal care and cosmetic formulations where surfactants are widely used can be quoted. The most widely used systems are perhaps hand creams and lotions. Both are formulated as oil-in-water (O/W) or water-in-oil (W/O) systems using surfactant mixtures which are used for the emulsification process as well as formation of liquid crystalline phases (mostly of the lamellar type) that wrap around the droplets and/or form "gel" networks in which the oil droplets are incorporated. These lamellar phases can be produced in emulsion systems by using a combination of surfactants with various HLB numbers and choosing the right oil (emollient). In many cases liposomes and vesicles are also produced by using lipids of various compositions. Two main types of lamellar liquid crystalline structures can be produced: "Oleosomes" and "Hydrosomes" (Fig. 11.1)

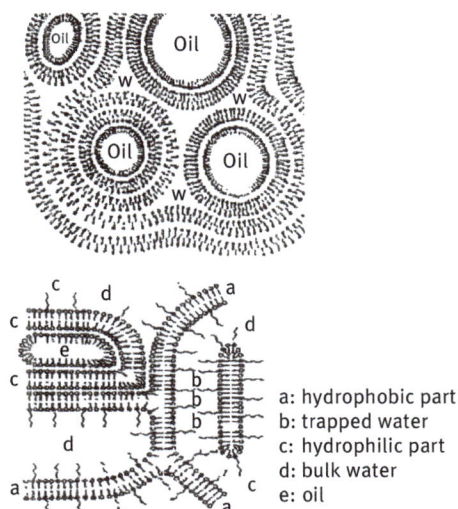

a: hydrophobic part
b: trapped water
c: hydrophilic part
d: bulk water
e: oil

Fig. 11.1. Schematic representation of "Oleosomes" and "Hydrosomes".

Several advantages of lamellar liquid crystalline phases in cosmetics can be quoted:
1. they produce an effective barrier against coalescence;
2. they can produce "gel networks" that provide the right consistency for application as well as prevention of creaming or sedimentation;
3. they can influence the delivery of active ingredients both of the lipophilic and hydrophilic types;
4. since they mimic the skin structure (in particular the stratum corneum) they can offer prolonged hydration potential.

The second and important class of emulsions is those covering the size range 20–200 nm (see Chapter 8). As mentioned in Chapter 8, the attraction of nanoemulsions for application in personal care and cosmetics is due to the following advantages:
1. The very small droplet size causes a large reduction in the gravity force and the Brownian motion may be sufficient for overcoming gravity. This means that no creaming or sedimentation occurs on storage.
2. The small droplet size also prevents any flocculation of the droplets. Weak flocculation is prevented and this enables the system to remain dispersed with no separation.
3. The small droplets also prevent their coalescence, since these droplets are non-deformable and hence surface fluctuations are prevented. In addition, the significant surfactant film thickness (relative to droplet radius) prevents any thinning or disruption of the liquid film between the droplets.

4. Nanoemulsions are suitable for efficient delivery of active ingredients through the skin. The large surface area of the emulsion system allows rapid penetration of actives.
5. Due to their small size, nanoemulsions can penetrate through the "rough" skin surface and this enhances penetration of actives.
6. The transparent nature of the system, their fluidity (at reasonable oil concentrations) as well as the absence of any thickeners may give them a pleasant aesthetic character and skin feel.
7. Unlike microemulsions (which require a high surfactant concentration, usually in the region of 20 % and higher), nanoemulsions can be prepared using reasonable surfactant concentration. For a 20 % O/W nanoemulsion, a surfactant concentration in the region of 5–10 % may be sufficient.
8. The small size of the droplets allows them to deposit uniformly on substrates. Wetting, spreading and penetration may be also enhanced as a result of the low surface tension of the whole system and the low interfacial tension of the O/W droplets.
9. Nanoemulsions can be applied for delivery of fragrants which may be incorporated in many personal care products. This could also be applied in perfumes which are desirable to be formulated alcohol free.
10. Nanoemulsions may be applied as a substitute for liposomes and vesicles (which are much less stable) and it is possible in some cases to build lamellar liquid crystalline phases around the nanoemulsion droplets.

A third class of emulsions that is used in personal care and cosmetics is the multiple emulsions. Multiple emulsions are complex systems of emulsions of emulsions: Water-in-Oil-in-Water (W/O/W); Oil-in-Water-in-Oil (O/W/O). The W/O/W multiple emulsions are the most commonly used systems in personal care products. Multiple emulsions are ideal systems for application in cosmetics:
1. one can dissolve actives in three different compartments;
2. they can be used for controlled and sustained release;
3. they can be applied as creams by using thickeners in the outer continuous phase.

Multiple emulsions are conveniently prepared by a two-step process. For W/O/W, a W/O emulsion is first prepared using a low HLB polymeric surfactant using a high speed stirrer to produce droplets ~1 µm. The W/O emulsion is then emulsified in an aqueous solution containing a high HLB polymeric surfactant using a low speed stirrer to produce droplets 10–100 µm.

To prepare a stable multiple emulsion, the following criteria must be satisfied:
1. two emulsifiers with low and high HLB numbers to produce the primary W/O emulsion and the final W/O/W multiple emulsion;
2. Polymeric emulsifiers that provide steric stabilization are necessary to maintain the long term physical stability;

3. optimum osmotic balance for W/O/W between the internal water droplets and outer continuous phase. This can be achieved by using electrolytes or non-electrolytes.

As mentioned above, multiple emulsions are conveniently prepared using a two-step process: A W/O system is first prepared by emulsification of the aqueous phase (which may contain an electrolyte to control the osmotic pressure) into an oil solution of the polymeric surfactant with low HLB number. A high speed stirrer is used to produce droplets ~ 1 μm. The droplet size of the primary emulsion can be determined using dynamic light scattering. The primary W/O emulsion is then emulsified into an aqueous solution (of an electrolyte to control the osmotic pressure) containing the polymeric surfactant with high HLB number. In this case a low speed stirrer is used to produce multiple emulsion droplets in the range 10–100 μm. The droplet size of the multiple emulsion can be determined using optical microscopy (with image analysis) or using light diffraction techniques (Malvern Mastersizer). A schematic representation of the preparation of W/O/W multiple emulsions is shown in Fig. 11.2.

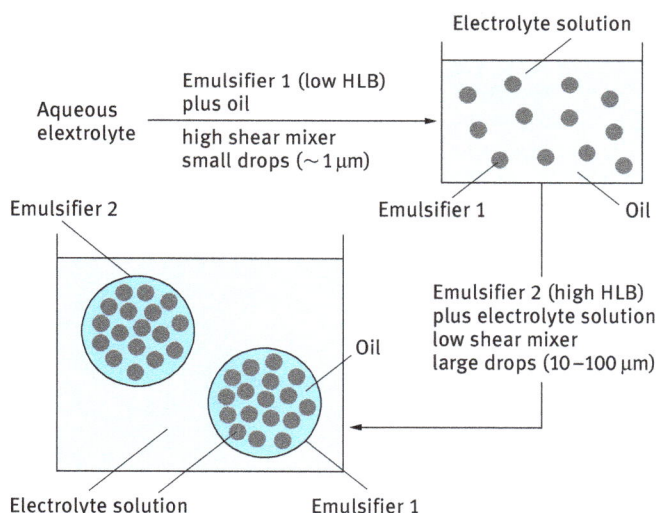

Fig. 11.2. Scheme for preparation of W/O/W multiple emulsion.

Another important application of surfactants in personal care and cosmetics is the preparation of liposomes and vesicles. Liposomes are multilamellar structures consisting of several bilayers of lipids (several μm) – they are produced by simply shaking an aqueous solution of phospholipids, e.g. egg lecithin. When sonicated, these multilayer structures produce unilamellar structures (with size range of 25–50 nm) that are referred to as liposomes. A schematic picture of liposomes

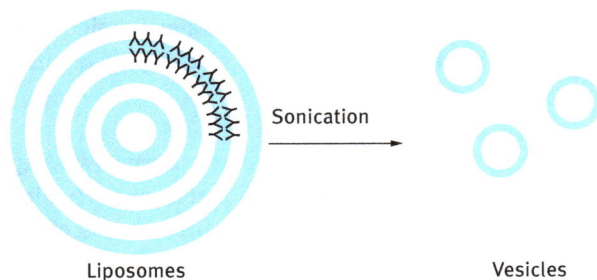

Fig. 11.3. Schematic representation of liposomes and vesicles.

and vesicles is given in Fig. 11.3. Glycerol-containing phospholipids are used for the preparation of liposomes and vesicles: phosphatidylcholine; phosphatidylserine; phosphatidylethanolamine; phosphatidylanisitol; phosphatidylglycerol; phosphatidic acid; cholesterol. In most preparations, a mixture of lipids is used to obtain the optimum structure. Liposomes and vesicles are ideal systems for cosmetic applications. They offer a convenient method for solubilizing non-polar active substances in the hydrocarbon core of the bilayer. Polar substances can also be intercalated in the aqueous layer between the bilayer. They will also form lamellar liquid crystalline phases and they do not disrupt the stratum corneum. No facilitated transdermal transport is possible thus eliminating skin irritation. Phospholipid liposomes can be used as in vitro indicators for studying skin irritation by surfactants.

Several other examples of personal care formulations, where surfactants are applied can be quoted and these are summarized below.

11.1.1 Shaving formulations

Three main types of shaving preparations may be distinguished:
1. wet shaving formulations;
2. dry shaving formulations and
3. after shave preparations.

The main requirements for a wet shaving preparation are to soften the beard, to lubricate the passage of the razor over the face and to support the beard hair. The hair of a typical beard is very coarse and difficult to cut and hence it is important to soften the hair for easier shaving and this requires the application of soap and water. The soap makes the hair hydrophilic and hence it becomes easy to wet by water which also may cause swelling of the hair. Most soaps used in shaving preparations are sodium or potassium salts of long chain fatty acids (sodium or potassium stearate or palmitate). Sometimes, the fatty acid is neutralized with triethanolamine. Other surfactants such as ether sulfates and sodium lauryl sulfate are included in the formulation to produce stable foam. Humectants such as glycerol may also be included to hold the moisture and prevent drying of the lather during shaving. Dry shaving is a process using electric shavers. In contrast to wet shaving, when using an electric razor the hair should remain dry and stiff. This requires removal of the moisture film and sebum from the face. This may be achieved by using a lotion based on an alcohol solution. A lubricant such as fatty acid ester or isopropyl myristate may be added to the lotion. Alternatively, a dry talc stick may be used that can absorb the moisture and sebum from the face. Another important formulation that is used after shaving is that used to reduce skin irritation and provides a pleasant feel. This can be achieved by providing emolliency accompanied by a cooling effect. In some cases an antiseptic agent is added to keep the skin free from bacterial infection. Most of these after shave formulations are aqueous based gels which should be non-greasy and easy to rub into the skin.

11.1.2 Bar soaps

These are one of the oldest toiletries products that have been used for centuries. The earliest formulations were based on simply fatty acid salts, such as sodium or potassium palmitate. However, these simple soaps suffer from the problem of calcium soap precipitation in hard water. For that reason, most soap bars contain other surfactants such as cocomonoglyceride sulfate or sodium cocoglyceryl ether sulfonate that prevent precipitation with calcium ions. Other surfactants used in soap bars include sodium cocyl isethinate, sodium dodecyl benzene sulfonate and sodium stearly sulfate.

11.1.3 Liquid hand soaps

Liquid hand soaps are concentrated surfactant solutions which can be simply applied from a plastic squeeze bottle or a simple pump container. The formulation consists of a mixture of various surfactants such as alpha olefin sulfonates, lauryl sulfates or lauryl ether sulfates. Foam boosters such as cocoamides are added to the formulation. A moisturizing agent such as glycerine is also added. A polymer such as poyquaternium-7 is added to hold the moisturizers and to impart a good skin feel. More recently, some manufacturers used alkyl polyglucosides in their formulations. The formulation may also contain other ingredients such as proteins, mineral oil, silicones, lanolin, etc. In many cases a fragrant is added to impart a pleasant smell to the liquid soap.

11.1.4 Bath oils

Three types of bath oils may be distinguished: floating or spreading oil, dispersible, emulsifying or blooming oil and milky oil. The floating or spreading bath oils (usually mineral or vegetable oils or cosmetic esters such as isopropyl myristate) are the most effective for lubricating dry skin as well as carrying the fragrant. However, they suffer from "greasiness" and deposit formation around the bath tub. These problems are overcome by using self-emulsifying oils which are formulated with surfactant mixtures. When added to water they spontaneously emulsify forming small oil droplets that deposit on the skin surface. However, these self-emulsifying oils produce less emolliency when compared with the floating oils. These bath oils usually contain a high level of fragrance since they are used in a large amount of water.

11.1.5 Foam (or bubble) baths

These can be produced in the form of liquids, creams, gels, powders, granules (beads). Their main function is to produce maximum foam into running water. The basic surfactant used in bubble bath formulations are anionic, nonionic or amphoteric together with some foam stabilizers, fragrants and suitable solubilizers. These formulations should be compatible with soap and they may contain other ingredients for enhancing skin care properties.

11.1.6 After bath preparations

These are formulations designed to counteract the damaging effects caused after bathing, e.g. skin drying caused by removal of natural fats and oils from the skin. Several formulations may be used, e.g. lotions and creams, liquid splashes, dry oil spray, dusting powders or talc, etc. The lotions and creams which are the most commonly used formulations are simply O/W emulsions with skin conditioners and emollients. The liquid splashes are hydroalcoholic products that contain some oil to provide skin conditioning. They can be applied as a liquid spread on the skin by hand or by spraying.

11.1.7 Skin care products

The skin forms an efficient permeability barrier with the following essential functions:
1. protection against physical injury, wear and tear and it may also protect against ultraviolet (uv) radiation;
2. it protects against penetration of noxious foreign materials including water and micro-organisms;

3. it controls loss of fluids, salts, hormones and other endogenous materials from within;
4. it provides thermoregulation of the body by water evaporation (through sweat glands).

For these reasons skin care products are essential materials for protection against skin damage. A skin care product should have two main ingredients, a moisturizer (humectant) that prevents water loss from the skin and an emollient (the oil phase in the formulation) that provides smoothing, spreading, degree of occlusion and moisturizing effect. The term emollient is sometimes used to encompass both humectant and oils. The moisturizer should keep the skin humid and it should bind moisture in the formulation (reducing water activity) and protect it from drying out. The term water content implies the total amount of water in the formulation (both free and bound), whereas water activity is a measure of the free (available) water only. The water content of the deeper, living epidermic layers is of the order of 70 % (same as the water content in living cells). Several factors can be considered to account for drying of the skin. One should distinguish between the water content of the dermis, viable epidermis and the horny layer (stratum corneum). During dermis aging, the amount of mucopolysaccharides decreases leading to a decrease in the water content. This aging process is accelerated by uv radiation (in particular the deep penetrating UVA; see section on sunscreens). Chemical or physical changes during aging of the epidermis also lead to dry skin. The structured lipid/water bilayer system in the stratum corneum forms a barrier towards water loss and protects the viable epidermis from the penetration of exogenous irritants. The skin barrier may be damaged by extraction of lipids by solvents or surfactants and the water loss can also be caused by low relative humidity. Dry skin, caused by a loss of horny layer can be cured by formulations containing extracts of lipids from horny layers of humans or animals. Due to loss of water from the lamellar liquid crystalline lipid bilayers of the horny layer, phase transition to crystalline structures may occur and this causes contraction of the intercellular regions. The dry skin becomes inflexible and inelastic and it may also crack.

For the above reasons, it is essential to use skin care formulations that contain moisturizers (e.g. glycerine) that draw and strongly bind water, thus trapping water on the skin surface. Formulations prepared with non-polar oils (e.g. paraffin oil) also help in water retention. Occlusion of oil droplets on the skin surface reduces the rate of trans-epidermal water loss. Several emollients can be applied, e.g. petrolatum, mineral oils, vegetable oils, lanolin and its substitutes and silicone fluids. Apart from glycerine, which is the most widely used humectant, several other moisturizers can be used, e.g. sorbitol, propylene glycol, polyethylene glycols (with molecular weights in the range 200–600). As mentioned above, liposomes or vesicles, and neosomes can also be used as skin moisturizers.

In general, emollients may be described as products that have softening and smoothing properties. They could be hydrophilic substances such as glycerine, sorbitol, etc. (mentioned above) and lipophilic oils such as paraffin oil, castor oil, triglyc-erides, etc. For the formulation of stable O/W or W/O emulsions for skin care ap-plications, the emulsifier system has to be chosen according to the polarity of the emollient. The polarity of an organic molecule may be described by its dielectric con-stant or dipole moment. Oil polarity can also be related to the interfacial tension of oil against water γ_{OW}. For example, a non-polar substance such as isopraffinic oil will give an interfacial tension in the region of 50 mNm^{-1}, whereas a polar oil such as cyclomethicone gives γ_{OW} in the region of 20 mNm^{-1}. The physicochemical nature of the oil phase determines its ability to spread on the skin, the degree of occlusivity and skin protection. The optimum emulsifier system also depends on the property of the oil (its HLB number) as discussed in detail in the chapter on emulsions.

The choice of an emollient for a skin care formulation is mostly based on sen-sorial evaluation using well-trained panels. These sensorial attributes are classified into several categories: ease of spreading, skin feeling directly after application and 10 minutes later, softness, etc. A lubricity test is also conducted to establish a fric-tion factor. Spreading of an emollient may also be evaluated by measurement of the spreading coefficient (see Chapter 10).

11.1.8 Hair care formulations

Hair care comprises two main operations:
1. care and stimulation of the metabolically active scalp tissue and its appendages, the pilosebaceous units. This process is normally carried out by dermatologists or specialized hair salons;
2. protection and care of the lifeless hair shaft as it passes beyond the surface of the skin. The latter is the subject of cosmetic preparations, which should acquire one or more of the following functions:
   (a) hair conditioning for ease of combing. This could also include formulations that can easily manage styling by combing and brushing and its capacity to stay in place for a while. The difficulty to manage hair is due to the static elec-tric charge which may be eliminated by hair conditioning;
   (b) hair "body", i.e. the apparent volume of a hair assembly as judged by sight and touch.

Another important type of cosmetic formulation is that used for hair dyeing, i.e. chang-ing the natural color of the hair. This subject will also be briefly discussed in this sec-tion. Hair is a complex multicomponent fiber with both hydrophilic and hydrophobic properties. It consists of 65–95 % by weight of protein and up to 32 % water, lipids, pigments and trace elements. The proteins are made of structured hard α-keratin em-bedded in an amorphous, proteinaceous matrix. Human hair is a modified epidermal structure taking its origin from small sacs called follicles that are located at the border line of dermis and hypodermis. A cross section of human hair shows three morphological regions, the medulla (inner core), the cortex that consists of fibrous proteins (α-keratin and amorphous protein) and an outer layer namely the cuticle. The major constituent of the cortex and cuticle of hair is protein or polypeptide (with several amino acid units). The keratin has an α-helix structure (molecular weight in the region of 40,000–70,000 daltons, i.e. 363–636 amino acid units).

The surface of hair has both acidic and basic groups (i.e. amphoteric in nature). For unaltered human hair, the maximum acid combining capacity is approximately 0.75 mmole/g hydrochloric, phosphoric or ethyl sulfuric acid. This value corresponds to the number of dibasic amino acid residues, i.e. arginine, lysine or histidine. The maximum alkali combining capacity for unaltered hair is 0.44 mmol/g potassium hydroxide. This value corresponds to the number of acidic residues, i.e. aspartic and glutamic side chains. The isoelectric point (i.e.p) of hair keratin (i.e. the pH at which there is an equal number of positive, $-NH^+$ and negative, $-COO^-$ groups) is ~pH = 6.0. However, for unaltered hair, the i.e.p is at pH = 3.67.

The above charges on human hair play an important role in the reaction of hair to cosmetic ingredients in a hair care formulation. Electrostatic interaction between anionic or cationic surfactants in any hair care formulation will occur with these charged groups. Another important factor in application of hair care products is the water content of the hair, which depends on the relative humidity (RH). At low RH (<25%), water is strongly bound to hydrophilic sites by hydrogen bonds (sometimes this is referred to as "immobile" water). At high RH (> 80%), the binding energy for water molecules is lower because of the multimolecular water-water interactions (this is sometimes referred to as "mobile" or "free" water). With increasing RH, the hair swells; increasing relative humidity from 0 to 100%, the hair diameter increases by ~14%. When water-soaked hair is put into a certain shape while drying, it will temporarily retain its shape. However, any change in RH may lead to the loss of setting.

Both surface and internal lipids exist in hair. The surface lipids are easily removed by shampooing with a formulation based on an anionic surfactant. Two successive steps are sufficient to remove the surface lipids. However, the internal lipids are difficult to remove by shampooing due to the slow penetration of surfactants.

Analysis of hair lipids reveals that they are very complex consisting of saturated and unsaturated, straight and branched fatty acids with chain length from 5 to 22 carbon atoms. The difference in composition of lipids between persons with "dry" and "oily" hair is only qualitative. Fine straight hair is more prone to "oiliness" than curly coarse hair.

From the above discussion, it is clear that hair treatment requires formulations for cleansing and conditioning of hair and this is mostly achieved by using shampoos. The latter are now widely used by most people and various commercial products are available with different claimed attributes. The primary function of a shampoo is to clean both hair and scalp of soils and dirt. Modern shampoos fulfill other purposes,

such as conditioning, dandruff control and sun protection. The main requirements for a hair shampoo are:
1. safe ingredients (low toxicity, low sensitization and low eye irritation);
2. low substantivity of the surfactants;
3. absence of ingredients that can damage the hair.

The main interactions of the surfactants and conditioners in the shampoo occur in the first few μm of the hair surface. Conditioning shampoos (sometimes referred to as 2-in-1 shampoos) deposit the conditioning agent onto the hair surface. These conditioners neutralize the charge on the surface of the hair, thus decreasing hair friction and this makes the hair easier to comb. The adsorption of the ingredients in a hair shampoo (surfactants and polymers) occurs both by electrostatic and hydrophobic forces. The hair surface has a negative charge at the pH at which a shampoo is formulated. Any positively charged species, such as a cationic surfactant or cationic polyelectrolye, will adsorb by electrostatic interaction between the negative groups on the hair surface and the positive head group of the surfactant. The adsorption of hydrophobic materials such as silicone or mineral oils occurs by hydrophobic interaction.

Several hair conditioners are used in shampoo formulations, e.g. cationic surfactants such as stearyl benzyl dimethyl ammonium chloride, cetyl trimethyl ammonium chloride, distearyl dimethyl ammonium chloride or stearamidopropyldimethyl amine. As mentioned above, these cationic surfactants cause dissipation of static charges on the hair surface, thus allowing ease of combing by decreasing the hair friction. Sometimes, long chain alcohols such as cetyl alcohol, stearyl alcohol and cetostearyl alcohol are added, which is claimed to have a synergistic effect on hair conditioning. Thickening agents, such as hydroxyethyl cellulose or xanthan gum are added, which act as rheology modifiers for the shampoo and may also enhance deposition to the hair surface. Most shampoos also contain lipophilic oils such as dimethicone or mineral oils, which are emulsified into the aqueous surfactant solution. Several other ingredients, such as fragrants, preservatives and proteins are also incorporated in the formulation. Thus, a formula of shampoo contains several ingredients and the interaction between the various components should be considered both for the long-term physical stability of the formulation and its efficiency in cleaning and conditioning the hair.

Another hair care formulation is that used for permanent-waving, straightening and depilation. The steps in hair waving involve reduction, shaping and hardening of the hair fibers. Reduction of cystine bonds (disulfide bonds) is the primary reaction in permanent waving, straightening and depilation of human hair. The most commonly used depilatory ingredient is calcium thioglycollate that is applied at pH 11–12. Urea is added to increase the swelling of the hair fibers. In permanent waving, this reduction is followed by molecular shifting through stressing the hair on rollers and ended by neutralization with an oxidizing agent where cysteine bonds are reformed. Recently, superior "cold waves" have replaced the "hot waves" by using thioglycollic acid at pH 9 to 9.5. Glycerylmonothio-glycolate is also used in hair waving. An alternative reducing agent is sulfite, which could be applied at pH 6 and this followed by hydrogen peroxide neutralizer.

Another process that is also applied in the cosmetic industry is hair bleaching which has the main purpose of lightening the hair. Hydrogen peroxide is used as the primary oxidizing agent and salts of persulfate are added as "accelerators". The system is applied at pH 9–11. The alkaline hydrogen peroxide produces disintegration of the melanin granules, which are the main source of hair color, with subsequent destruction of the chromophore. Heavy metal complexants are added to reduce the rate of decomposition of the hydrogen peroxide. It should be mentioned that during hair bleaching, an attack of the hair keratin occurs producing cystic acid.

Another important formulation in the cosmetic industry is that used for hair dyeing. Three main steps may be involved in this process: bleaching, bleaching and coloring combined, as well as dyeing with artificial colors. Hair dyes can be classified into several categories: permanent or oxidative dyes, semipermanent dyes and temporary dyes or color rinses. The coloring agent for hair dyes may consist of an oxidative dye, an ionic dye, a metallic dye or a reactive dye. The permanent or oxidative dyes are the most commercially important systems and they consist of dye precursors such as p-phenylenediamine which is oxidized by hydrogen peroxide to a diimminium ion. The active intermediate condenses in the hair fiber with an electron-rich dye coupler such as resorcinol and with possibly electron-rich side chain groups of the hair, forming di-, tri- or polynuclear product that is oxidized into an indo dye.

Semipermanent dyes refer to formulations that dye the hair without the use of hydrogen peroxide to a color that only persists after 4–6 shampooings. The objective of temporary hair dyes or color rinses is to provide color that is removed after the first shampooing process.

11.1.9 Sunscreens

The damaging effect of sunlight (in particular ultraviolet light) has been recognized for several decades and this led to a significant demand for improved photoprotection by topical application of sunscreening agents. Three main wavelength of ultraviolet (UV) radiation may be distinguished, referred to as UV–A (wavelength 320–400, sometimes subdivided into UV–A1 (340–360) and UV-A2 (320–340)), UV–B (covering the wavelength 290–320) and UV–C (covering the wavelength range 200–290). The UV–C is of little practical importance since it is absorbed by the ozone layer of the stratosphere. The UV–B is energy rich and it produces intense short-range and long-range pathophysiological damage to the skin (sunburn). About 70 % is reflected by the horny layer (stratum corneum), 20 % penetrates into the deeper layers of the epidermis and 10 % reaches the dermis. UV–A is of lower energy, but its photobiological effects are cumulative causing long-term effects. UV–A penetrates deeply into the dermis and beyond, i.e., 20–30 % reaches the dermis. As it has a photoaugmenting effect on UV-B, it contributes about 8 % to UV–B erythema.

Several studies have shown that sunscreens are able not only to protect against UV-induced erythema in human and animal skin, but also to inhibit photocarcenogesis in animal skin. The increasing harmful effect of UV–A on UV–B has led to a quest for sunscreens that absorb the UV–A with the aim of reducing the direct dermal effects of UV–A which causes skin ageing and several other photosensitivity reactions. Sunscreens are given a sun protection factor (SPF) which is a measure of the ability of a sunscreen to protect against sunburn within the UV–B wavelength (290–320). The formulation of sunscreen with high SPF (> 50) has been the object of many cosmetic companies.

An ideal sunscreen formulation should protect against both UV–B and UV–A. Repeated exposure to UV–B accelerates skin ageing and can lead to skin cancer. UV–B can cause thickening of the horny layer (producing "thick" skin). UV–B can also cause damage to DNA and RNA. Individuals with fair skin cannot develop a protective tan and they must protect themselves from UV–B.

UV–A can cause also several effects:

1. large amounts of UV–A radiation penetrate deep into the skin and reach the dermis causing damage to blood vessels, collagen and elastic fibers;
2. prolonged exposure to UV–A can cause skin inflammation and erythema;
3. UV–A contributes to photoageing and skin cancer. It augments the biological effect of UV–B;
4. UV–A can cause phytotoxicity and photoallergy and it may cause immediate pigment darkening (immediate tanning) which may be undesirable for some ethnic populations.

From the above discussion, it is clear that formulation of effective sunscreen agents is necessary with the following requirements:

1. maximum absorption in the UV–B and/or UV–A;
2. high effectiveness at low dosage;
3. non-volatile agents with chemical and physical stability;
4. compatibility with other ingredients in the formulation;
5. sufficiently soluble or dispersible in cosmetic oils, emollients or in the water phase;
6. absence of any dermato-toxological effects with minimum skin penetration;
7. resistant to removal by perspiration.

Sunscreen agents may be classified into organic light filters of synthetic or natural origin and barrier substances or physical sunscreen agents. Examples of UV–B filters are cinnamates, benzophenones, p-aminobenzoic acid, salicylates, camphor derivatives and phenyl benzimidazosulphonates. Examples of UV–A filters are dibenzoyl methanes, anthranilates and camphor derivatives. Several natural sunscreen agents are available, e.g. camomile or aleo extracts, caffeic acid, unsaturated vegetable or animal oils. However, these natural sunscreen agents are less effective and they are seldom used in practice.

The barrier substances or physical sunscreens are essentially micronized insoluble organic molecules such as gaunine or micronized inorganic pigments such as titanium dioxide and zinc oxide. Micropigments act by reflection, diffraction and/or absorption of UV radiation. Maximum reflection occurs when the particle size of the pigment is about half the wave length of the radiation. Thus, for maximum reflection of UV radiation, the particle radius should be in the region of 140 to 200 nm. The uncoated materials such as titanium and zinc oxide can catalyze the photodecomposition of cosmetic ingredients such as sunscreens, vitamins, antioxidants and fragrances. These problems can be overcome by special coating or surface treatment of the oxide particles, e.g. using aluminum stearate, lecithins, fatty acids, silicones and other inorganic pigments. Most of these pigments are supplied as dispersions ready to mix in the cosmetic formulation. However, one must avoid any flocculation of the pigment particles or interaction with other ingredients in the formulation which causes severe reduction in their sunscreening effect.

A topical sunscreen product is formulated by the incorporation of one or more sunscreen agents (referred to as UV filters) in an appropriate vehicle, mostly an O/W or W/O emulsion. Several other formulations are also produced, e.g. gels, sticks, mousse (foam), spray formulation or an anhydrous ointment. In addition to the usual requirements for a cosmetic formulation, e.g. ease of application, pleasant aspect, color or touch, sunscreen formulations should also have the following characteristics:
1. effective in thin films, strongly absorbing both UV–B and UV–A;
2. non-penetrating and easily spreading on application;
3. should possess a moisturizing action and be waterproof and sweat resistant;
4. free from any phototoxic and allergic effect.

The majority of sunscreens on the market are creams or lotions (milks) and progress has been achieved in recent years to provide high SPF at low levels of sunscreen agents.

11.1.10 Make-up products

Make-up products include many systems such as lipstick, lip color, foundations, nail polish, mascara, etc. All these products contain a coloring agent which could be a soluble dye or a pigment (organic or inorganic). Examples of organic pigments are red, yellow, orange and blue lakes. The inorganic pigments comprise titanium dioxide, mica, zinc oxide, talc, iron oxide (red, yellow and black), ultramarines, chromium oxide, etc. Most pigments are modified by surface treatment using amino acids, chitin, lecithin, metal soaps, natural wax, polyacrylates, polyethylene, silicones, etc.

The color cosmetics comprise foundation, blushers, mascara, eyeliner, eye shadow, lip color and nail enamel. Their main function is to improve appearance, impart color, even out skin tones, hide imperfections and produce some protection. Several types of formulations are produced ranging from aqueous and nonaqueous suspensions to oil-in-water and water-in-oil emulsions and powders (pressed or loose).

The make-up products have to satisfy a number of criteria for acceptance by the consumer:
1. improved, wetting spreading and adhesion of the color components;
2. excellent skin feel;
3. skin and UV protection and absence of any skin irritation.

For these purposes, the formulation has to be optimized to achieve the desirable property. This is achieved by using surfactants and polymers as well as using modified pigments (by surface treatment). The particle size and shape of the pigments should also be optimized for proper skin feel and adhesion.

The pressed powders require special attention to achieve good skin feel and adhesion. The fillers and pigments have to be surface treated to achieve these objectives. Binders and compression aids are also added to obtain a suitable pressed powder. These binders can be dry powders, liquids or waxes. Other ingredients that may be added are sunscreens and preservatives. These pressed powders are applied in a simple way by simple "pick-up", deposition and even coverage. The appearance of the pressed powder film is very important and great care should be taken to achieve uniformity in an application. A typical pressed powder may contain 40–80 % fillers, 10–40 specialized fillers, 0–5 % binders, 5–10 % colorants, 0–10 % pearls and 3–8 % wet binders.

An alternative to pressed powders, liquid foundations have attracted special attention in recent years. Most of the foundation make-ups are made of O/W or W/O emulsions in which the pigments are dispersed either in the aqueous or the oil phase. These are complex systems consisting of a suspension/emulsion (suspoemulsion) formulation. Special attention should be made to the stability of the emulsion (absence of flocculation or coalescence) and suspension (absence of flocculation). This is achieved by using specialized surfactant systems such silicone polyols, block copolymers of poly(ethylene oxide) and poly(propylene oxide). Some thickeners may be also added to control the consistency (rheology) of the formulation.

The main purpose of a foundation make-up is to provide color in an even way, even out any skin tones and minimize the appearance of any imperfections. Humectants are also added to provide a moisturizing effect. The oil used should be chosen to be a good emollient. Wetting agents are also added to achieve good spreading and even coverage. The oil phase could be a mineral oil, an ester such as isopropyl myristate or volatile silicone oil (e.g. cyclomethicone). An emulsifier system of fatty acid/nonionic surfactant mixture may be used. The aqueous phase contains a humectant of glycerine, propylene glycol or polyethylene glycol. Wetting agents such as lecithin, low

HLB surfactant or phosphate esters may also be added. A high HLB surfactant may also be included in the aqueous phase to provide better stability when combined with the oil emulsifier system. Several suspending agents (thickeners) may be used such as magnesium aluminum silicate, cellulose gum, xanthan gum, hydroxyethyl cellulose or hydrophobically modified polyethylene oxide. A preservative such as methyl paraben is also included. The surface treated pigments are dispersed either in the oil or aqueous phase. Other additives such as fragrances, vitamins, and light diffusers may also be incorporated.

It is clear from the above discussion that liquid foundations represent a challenge to the formulation chemist due to the large number of components used and the interaction between the various components. Particular attention should be made to the interaction between the emulsion droplets and pigment particles (a phenomenon referred to as heteroflocculation) which may have adverse effects on the final property of the deposited film on the skin. Even coverage is the most desirable property and the optical property of the film, e.g. its light reflection, adsorption and scattering play important roles in the final appearance of the foundation film.

Several anhydrous liquid (or "semi-solid") foundations are also marketed by cosmetic companies. These may be described as cream powders consisting of a high content of pigment/fillers (40–50 %), a low HLB wetting agent (such as polysorbate 85), an emollient such as dimethicone combined with liquid fatty alcohols and some esters (e.g. octyl palmitate). Some waxes, such as stearyl dimethiicone or microcrystalline or carnuba wax, are also included in the formulation.

One of the most important make-up systems are lipsticks, which may be simply formulated with a pure fat base having a high gloss and excellent hiding power. However, these simple lipsticks tend to come off the skin too easily. In recent years, there was a great tendency to produce more "permanent" lipsticks which contain hydrophilic solvents such as glycols or tetrahydrfurfuryl alcohol. The raw materials for a lipstick base include: ozocerite (good oil absorbent that also prevents crystallization), microcrystalline ceresin wax (which also is a good oil absorbent), vaseline (that forms an impermeable film), beeswax (that increases resistance to fracture), myristyl myristate (that improves transfer to the skin), cetyl and meristyl lactate (that form an emulsion with moisture on the lip and is non-sticky), carnuba wax (an oil binder that increases the melting point of the base and gives some surface luster), lanolin derivatives, olyl alcohol and isopropyl myristate. This shows the complex nature of a lipstick base and several modifications of the base can produce some desirable effects that help good marketing of the product.

Mascara and eyeliners are also complex formulations that need to be carefully applied to the eye lashes and edges. Some of the preferred criteria for mascara are good deposition, ease of separation and lash curling. The appearance of the mascara should be as natural as possible. Lash lengthening and thickening are also desirable. The product should also remain for an adequate time and it should also be easily removable. Three types of formulations may be distinguished: anhydrous solvent-based suspension, water-in-oil emulsion and oil-in-water emulsion. Water resistance can be achieved by addition of emulsion polymers, e.g. polyvinyl acetate.

11.2 Surfactants in pharmacy [1, 3]

Surfactants are used in all disperse systems used in pharmaceutical formulations. Several types of disperse systems can be identified can be identified in pharmacy of which suspensions, emulsions and gels are the most commonly used. These disperse systems cover a wide size range: colloidal (1 nm–1 µm) and non-colloidal (> 1 µm). Several classes of surfactants are used in pharmacy:

1. Anionic surfactants such as alkali metal soaps, RCOOX, where X is sodium, potassium or ammonium. R is generally between C_{10} and C_{20}.
2. Sulfated fatty alcohols which are esters of sulfuric acid, the most commonly used compound is sodium lauryl sulfate, which is a mixture of sodium alkyl sulfates. The main component is sodium dodecyl sulfate, $C_{12}H_{25}-O-SO_3^-Na^+$. It is used pharmaceutically as preoperative skin cleanser having bacteriostatic action against gram-positive. It is also used in medicated shampoos and toothpaste (as foam producer).
3. Ether sulfates (sulfated polyoxyethylated alcohols). R–(OCH2–CH2)n–O–SO3–M+ (n < 6). This has better water solubility than the alkyl sulfates, better resistance to electrolyte and less irritation to the eye and the skin.
4. Sulfated oils, e.g. sulfated castor oil (triglyceride of the fatty acid 12-hydroxyoleic acid). This is used as an emulsifying agent for oil-in-water creams and ointments (non-irritant).
5. Cationic surfactants such as Cetrimide B. P., a mixture consisting of tetradecyl (∼ 68 %), dodecyl (∼22 %) and hexadecyl (∼ 7 %) trimethyl ammonium bromide. Solutions containing 0.1–1 % Cetrimide are used for cleansing skin, wounds and burns; also in shampoos to remove scales of seborrhoea; also in Cetavlon cream.
6. Banzalkonium chloride, a mixture of alkyl benzylammonium chlorides. In dilute solutions (0.1–0.2 %) it is used for pre-operative disinfection of the skin and mucous membranes, as a preservative for eye-drops.
7. Zwitterionic surfactants such as lecithin (phosphatidylcholine) which is applied as an oil-in-water emulsifier.
8. Nonionic surfactants which have the advantage over ionic surfactants in their compatibility with most other types of surfactants, little affected by moderate pH changes and moderate electrolyte concentrations. A useful scale for describing nonionic surfactants is the hydrophilic-lipophilic balance (HLB) which simply gives the relative proportion of hydrophilic to lipophilic components. For a simple nonionic surfactant such as an alcohol ethoxylate, the HLB is simply given by the percentage of hydrophilic components (PEO) divided by 5.

9. Sorbitan esters, which are mixtures of the partial esters of sorbitol and its mono- and di-anhydrides.
10. Polysorbates, which are the ethoxylated derivatives of the sorbitan esters. Commercial products are complex mixtures of partial esters of sorbitol and its mono- and di-anhydrides condensed with an approximate number of moles of ethylene oxide. They have high HLB numbers, are water soluble and are used as oil-in-water emulsifiers.
11. Polyoxyethylated glycol monoethers. These have the general structure $C_x E_y$, where x and y denote the alkyl and ethylene oxide chain length, e.g. $C_{12} E_6$ represents hexaoxyethylene glycol monododecyl ether. One of the most widely used compounds is Cetomacrogel 1000 BPC, which is a water soluble compound with an alkyl chain length of 15 or 17 and an ethylene oxide chain length between 20 and 24. It is used in the form of cetomacrogel emulsifying wax in the preparation of oil-in-water emulsions and also as a solubilizing agent for volatile oils.
12. Polymeric surfactants; the most commonly used polymeric surfactants in pharmacy are the A–B–A block copolymers, with A being the hydrophilic chain (polyethylene oxide, PEO) and B being the hydrophobic chain (polypropylene oxide, PPO). The general structure is PEO–PPO–PEO and is commercially available with different proportions of PEO and PPO (Pluronics or Poloxamers). The commercial name is followed by a letter L (Liquid), P (Paste) and F (Flake). This is followed by two numbers that represent the composition – the first digit represents the PPO molecular mass and the second digit represents the % of PEO: Pluronic F68 (PPO mol wt 1501–1800) + 140 mol EO; Pluronic L62 (PPO mol tt 1501–1800) + 15 mol EO.

11.2.1 Surface active drugs

A large number of drugs are surface active, e.g. chlorpromazine, diphenylmethane derivatives (such as diphenhydramine) and tricyclic antidepressants (such as amitriptyline). The solution properties of these surface active drugs and their mode of association play an important role in their biological efficacy. Many drugs exhibit surface active properties that are similar to surfactants, e.g. they accumulate at interfaces and produce aggregates (micelles) at critical concentrations. However, micellization of drugs represent only one pattern of association, since with many drug molecules rigid aromatic or heterocyclic chains replace the flexible hydrophobic chains present in most surfactant systems. This will have a pronounced effect on the mode of association, to an extent that the process may not be regarded as micellization. A self-association structure may be produced by hydrophobic interaction (charge repulsion plays an insignificant role in this case) and the process is generally continuous, i.e. with no abrupt change in the properties. It should be mentioned, however, that many drug molecules may contain aromatic groups with a high degree of flexibility. In this case, the association structures resemble surfactant micelles. However, the aggregation numbers of these association units are much lower (in the region of 9–12) than those encountered with micellar surfactants (which show aggregation numbers of 50 or more depending on the alkyl chain length). This lower aggregation number cast some doubt on micelle formation and a continuous association process may be envisaged instead.

Both the surface activity and micellization have implications on the biological efficacy of many drugs. Surface active drugs tend to bind hydrophobically to proteins and other biological macromolecules. They also tend to associate with other amphipathic molecules such as other drugs, bile salts and of course with receptors. The activity of phenothiazines is attributed to their interaction with membranes, which may be correlated with their surface activity. It is believed that these compounds act by altering the conformation and activity of enzymes and by altering membrane permeability and function.

Several other examples may be quoted to illustrate the importance of surface activity of many drugs. Many drugs produce intralysosomal accumulation of phospholipids which are observable as multilamellar objects within the cell. The drugs which are implicated in phospholipidosis induction are often amphipathic compounds. The interaction between the surfactant drug molecules and phospholipid render the phospholipid resistant to degradation by lysosomal enzymes resulting in their accumulation in cells.

Many local anesthetics have significant surface activity and it is tempting to correlate their surface activity to their action. However, one should not forget other important factors such as partitioning of the drug into the nerve membrane (a factor that depends on the pK_a) and the distribution of hydrophobic and cationic groups which must be important for the appropriate disruption of nerve membrane function.

The biological relevance of micelle formation by drug molecules is not as clear as their surface activity, since the drug is usually applied at concentration well below that at which micelles are formed. However, accumulation of drug molecules in certain sites may allow them to reach concentrations whereby micelles are produced. Such aggregate units may cause significant biological effects. For example, the concentration of monomeric species may increase only slowly or may decrease with increase in total concentration and the transport and colligative properties of the system are changed. In other words, the aggregation of the compounds will affect their thermodynamic activity and hence their biological efficacy in vivo.

11.2.2 Naturally occurring micelle-forming systems

Several naturally occurring amphipathic molecules (in the body) exist, such as bile salts, phospholipids, and cholesterol, play an important role in various biological processes. Their interactions with other solutes, such as drug molecules, and with membranes are also very important. Bile salts are synthesized in the liver and they consist of alicyclic compounds possessing hydroxyl and carboxyl groups. It is the positioning of the hydrophilic groups in relation to the hydrophobic steroidal nucleus that gives the bile salts their surface activity and determines their ability to aggregate. It has been suggested that small or primary aggregates with up to 10 monomers form above the cmc by hydrophobic interactions between the non-polar side of the monomers. These primary aggregates form larger units by hydrogen bonding between the primary micelles.

The cmc of bile salts is strongly influenced by its structure; the trihydroxy cholanic acids have higher cmc than the less hydrophilic dihydroxy derivatives. As expected, the pH of solutions of these carboxylic acid salts has an influence on micelle formation. At sufficiently low pH, bile acids that are sparingly soluble will be precipitated from solution, initially being incorporated or solubilized in the existing micelles. The pH at which precipitation occurs, on saturation of the micellar system, is generally about one pH unit higher than the pK_a of the bile acid.

Bile salts play important roles in physiological functions and drug absorption. It is generally agreed that bile salts aid fat absorption. Mixed micelles of bile salts, fatty acids and monogylcerides can act as vehicles for fat transport. However, the role of bile salts in drug transport is not well understood. Several suggestions have been made to explain the role of bile salts in drug transport, such as facilitation of transport from liver to bile by direct effect on canicular membranes, stimulation of micelle formation inside the liver cells, binding of drug anions to micelles, etc. The enhanced absorption of medicinals on administration with deoxycholic acid may be due to a reduction in interfacial tension or in micelle formation. The administration of quinine and other alkaloids in combination with bile salts has been claimed to enhance their parasiticidal action. Quinine, taken orally, is considered to be absorbed mainly from the intestine and a considerable amount of bile salts is required to maintain a colloidal dispersion of quinine. Bile salts may also influence drug absorption either by affecting membrane permeability or by altering normal gastric emptying rates. For example, sodium taurcholate increases the absorption of sulfaguanidine from the stomach, jejunum and ileum. This is due to increasing membrane permeability induced by calcium depletion and interference with the bonding between phospholipids in the membrane.

Another important naturally occurring class of surfactants which are widely found in biological membranes are the lipids, such as phosphatidylcholine (lecithin), lysolecithin, phosphatidylethanolamine and phospahitidyl inositol. These lipids are also used as emulsifiers for intravenous fat emulsions, anesthetic emulsions as well as for production of liposomes or vesicles for drug delivery. The lipids form coarse turbid dispersions of large aggregates (liposomes) which on ultrasonic irradiation form smaller units or vesicles. The liposomes are smectic mesophases of phospholipids organized into bilayers which assume a multilamellar or unilamellar structure. The multilamellar species are heterogeneous aggregates, most commonly prepared by dispersal of a thin film of phospholipid (alone or with cholesterol) into water. Sonication of the multilamellar units can produce the unilamellar liposomes, sometimes referred to as vesicles. The net charge of liposomes can be varied by incorporation of a long chain amine, such as stearyl amine (to give a positively charged vesicle) or dicetyl phosphate (giving negatively charged species). Both lipid-soluble and water-soluble drugs can be entrapped in liposomes. The liposoluble drugs are solubilized in the hydrocarbon interiors of the lipid bilayers, whereas the water-soluble drugs are intercalated in the aqueous layers. Liposomes, like micelles, may provide a special medium for reactions to occur between the molecules intercalated in the lipid bilayers or between the molecules entrapped in the vesicle and free solute molecules.

Phospholipids play an important role in lung functions. The surface active material to be found in the alveolar lining of the lung is a mixture of phospholipids, neutral lipids and proteins. The lowering of surface tension by the lung surfactant system and the surface elasticity of the surface layers assists alveolar expansion and contraction. Deficiency of lung surfactants in newborns leads to a respiratory distress syndrome and this led to the suggestion that instillation of phospholipid surfactants could cure the problem.

11.2.3 Biological implications of the presence of surfactants in pharmaceutical formulations

The use of surfactants as emulsifying agents, solubilizers, dispersants for suspensions and as wetting agents in a formulation can lead to significant changes in the biological activity of the drug in the formulation. Surfactant molecules incorporated in the formulation can affect drug availability and its interaction with various sites in several ways. The surfactant may influence the desegregation and dissolution of solid dosage forms, by controlling the rate of precipitation of drugs administered in solution form, by increasing membrane permeability and affecting membrane integrity. Release of poorly soluble drugs from tablets and capsules for oral use may be increased by the presence of surfactants, which may decrease the aggregation of drug particles and, therefore, increase the area of the particles available for dissolution. The lowering of surface tension may also be a factor in aiding the penetration of water into the drug mass. This wetting effect operates at low surfactant concentration. Above the cmc, the increase in saturation solubility of the drug substance by solubilization in the surfactant micelles can result in more rapid rates of drug dissolution. This will increase the rate of drug entry into the blood and may affect peak blood levels. However, very high concentrations of surfactant can decrease drug absorption by decreasing the chemical potential of the drug. This results when the surfactant concentration exceeds that required to solubilize the drug. Complex interactions between the surfactants and proteins may take place and this will result in alteration of drug metabolizing enzyme activity. There have also been some suggestions that the surfactant may influence the binding of the drug to the receptor site. Some surfactants have direct physiological activity of their own and in the whole body these molecules can affect the physiological environment, e.g. by altering gastric residence time.

Numerous studies on the influence of surfactants on drug absorption have shown them to be capable of increasing, decreasing or exerting no effect on the transfer of drugs through membranes. As discussed above, the presence of surfactants affects the dissolution rate of the drug.

11.2.4 Solubilized systems

Solubilization is the process of preparation of thermodynamically stable isotropic solution of a substance (normally insoluble or sparingly soluble in a given solvent) by incorporation of an additional amphiphilic component(s). It is the incorporation of the compound (referred to as solubilizate or substrate) within micellar (L_1 phase) or reverse micellar (L_2 phase) system. Several factors affect solubilization:

1. Solubilizate structure: generalizations about the manner in which structure affects solubilization are complicated by the existence of different solubilization sites. The main parameters that may be considered when investigating solubilizates are: polarity, polarizability, chain length and branching, molecular size and shape. The most significant effect is perhaps the polarity of the solubilizate and sometimes they are classified into polar and apolar; however, difficulty exists with intermediate compounds. Some correlation exists between hydrophilicity/lipophilicity of solubilizate and partition coefficient between octanol and water (the log P number concept, the higher the value the more lipophilic the compound is).

2. Surfactant structure: for solubilizates incorporated in the hydrocarbon core, the extent of solubilization increases with increase in the alkyl chain length. For the same R, solubilization increases in the order: anionics <cationics <nonionics. The solubilization power that is normally described by the ratio of moles solubilizate to mole surfactant increases with increase in the PEO chain length. This is due to the decrease in micelle size. With increasing the PEO chain length, the aggregation number decreases and hence the number of micelles per mole surfactant increases.

3. Temperature: mostly solubilization increases with increasing temperature as a result of the increase in solubility of the compound and decrease of the cmc (for nonionic surfactants) with increase of temperature.

4. Addition of electrolytes and non-electrolytes: most electrolytes cause a reduction in the cmc and they may increase the aggregation number (and size) of the micelle. This may lead to an increase in solubilization. Addition of non-electrolytes, e.g. alcohols can lead to an increase in solubilization.

The above discussion clearly demonstrates that solubilization above the cmc offers an approach to formulation of poorly soluble drugs. This approach has several limitations: finite capacity of micelles for the drug; short- or long-term adverse effects; solubilization of other ingredients such as preservatives, flavors and coloring agents, which may cause alteration in stability and effectiveness.

11.2.5 Pharmaceutical aspects of solubilization

The presence of micelles and surfactant monomers in a drug formulation can have pronounced effects on the biological efficacy. Surfactants (both micelles and monomers) can influence the disintegration and dissolution of solid dosage forms by controlling the rate of precipitation (drug administration in solution), increasing membrane permeability and affecting membrane integrity. The release of poorly soluble drugs from tablets and capsules (oral use) may be increased in the presence of surfactants. The reduction of aggregation on disintegration of tablets and capsules increases the surface area. Lowering of surface tension aids penetration of water into the drug mass. Above the cmc, increasing flux by solubilization can lead to a rapid increase in the rate of dissolution. However, very high surfactant concentrations (above that required for solubilization) may decrease drug absorption by decreasing the chemical potential of the drug. The complex interaction between surfactant micelles, monomers and proteins may alter the drug metabolizing activity. Surfactants may also alter the binding of drug to receptor site.

11.3 Surfactants in agrochemicals [1, 4, 5]

The formulations of agrochemicals cover a wide range of systems that are prepared to suit a specific application. Several types can be quoted of which the following are the most important: emulsifiable concentrates (EC's), emulsions (EW's), suspension concentrates (SC's). suspoemulsions (mixtures of suspensions and emulsions), microemulsions and capsules (controlled release formulations). All these formulations require the use of a surfactant, which is not only essential for their preparation and maintenance of their long-term physical stability, but also for enhancement of the biological performance of the agrochemical.

Several surfactant types are used in agrochemical formulations of which the anionics are probably the most commonly used. This is due to their relatively low cost of manufacture and they are practically used in every type of formulation. Linear chains are preferred since they are more effective and more degradable than the branched chains. The most commonly used hydrophilic groups are carboxylates, sulfates, sulfonates and phosphates. A general formula may be ascribed to anionic surfactants as follows:

Carboxylates: $\quad C_nH_{2n+1}\,COO^-\,X$
Sulfates: $\quad\quad\ C_nH_{2n+1}\,OSO_3^-\,X$
Sulfonates: $\quad\ \ C_nH_{2n+1}\,SO_3^-\,X$
Phosphates: $\quad\ C_nH_{2n+1}\,OPO(OH)O^-\,X$

with n being the range 8–16 atoms and the counterion X is usually Na^+.

Several other anionic surfactants are commercially available such as sulfosucci-nates, isethionates and taurates and these are sometimes used for special applications. Phosphate-containing anionic surfactants are also used in some applications. Both alkyl phosphates and alkyl ether phosphates are made by treating the fatty alcohol or alcohol ethoxylates with a phophorylating agent, usually phosphorous pentoxide, P_4O_{10}. The reaction yields a mixture of mono- and diesters of phosphoric acid. The ratio of the two esters is determined by the ratio of the reactants and the amount of water present in the reaction mixture. The physicochemical properties of the alkyl phosphate surfactants depend on the ratio of the esters.

The most common cationic surfactants used in agrochemical formulations are the quaternary ammonium compounds with the general formula $R'R''R'''R''''N^+X^-$, where X^- is usually a chloride ion and R represents alkyl groups. A common class of cationics is the alkyl trimethyl ammonium chloride, where R contains 8–18 C atoms, e.g. dodecyl trimethyl ammonium chloride, $C_{12}H_{25}(CH_3)_3NCl$. Another cationic surfactant class is that containing two long chain alkyl groups, i.e. dialkyl dimethyl ammonium chloride, with the alkyl groups having a chain length of 8–18 C atoms. These dialkyl surfactants are less soluble in water than the monoalkyl quaternary compounds, but they are sometimes used in agrochemical formulations as adjuvants and/or rheology modifiers. A special cationic surfactant is alkyl dimethyl benzyl ammonium chloride (sometimes referred to as benzalkonium chloride), which may be also used in some formulations as an adjuvant. Imidazolines can also form quaternaries, the most common product being the ditallow derivative quaternized with dimethyl sulfate. Cationic surfactants can also be modified by incorporating polyethylene oxide chains, e.g. dodecyl methyl polyethylene oxide ammonium chloride. Cationic surfactants are generally water soluble when there is only one long alkyl group. They are generally compatible with most inorganic ions and hard water. Cationics are generally stable to pH changes, both acid and alkaline. They are incompatible with most anionic surfactants, but they are compatible with nonionics. These cationic surfactants are insoluble in hydrocarbon oils. In contrast, cationics with two or more long alkyl chains are soluble in hydrocarbon solvents, but they become only dispersible in water (sometimes forming bilayer vesicle type structures). They are generally chemically stable and can tolerate electrolytes. The cmc of cationic surfactants is close to that of anionics with the same alkyl chain length.

Amphoteric surfactants containing both cationic and anionic groups are also used in some formulations. The most common amphoterics are the N-alkyl betaines which are derivatives of trimethyl glycine $(CH_3)_3NCH_2COOH$ (that was described as betaine). An example of a betaine surfactant is lauryl amido propyl dimethyl betaine $C_{12}H_{25}CON(CH_3)_2CH_2COOH$. These alkyl betaines are sometimes described as alkyl dimethyl glycinates.

The main characteristics of amphoteric surfactants are their dependence on the pH of the solution in which they are dissolved. In acid pH solutions, the molecule acquires a positive charge and it behaves like a cationic, whereas in alkaline pH solutions, it becomes negatively charged and behaves like an anionic. A specific pH can be defined at which both ionic groups show equal ionization (the isoelectric point of the molecule). This can be described by the following scheme:

$$N^+ \ldots COOH \quad \leftrightarrow \quad N^+ \ldots COO^- \quad \leftrightarrow \quad NH \ldots COO^-$$

| acid pH < 3 | isoelectric pH > 6 | alkaline |

Amphoteric surfactants are sometimes referred to as zwitterionic molecules. They are soluble in water, but the solubility shows a minimum at the isoelectric point. Amphoterics show excellent compatibility with other surfactants, forming mixed micelles. They are chemically stable both in acids and alkalis. The surface activity of amphoterics varies widely and it depends on the distance between the charged groups that show a maximum in surface activity at the isoelectric point.

The most common nonionic surfactants are those based on ethylene oxide, referred to as ethoxylated surfactants. Several classes can be distinguished: alcohol ethoxylates, alkyl phenol ethoxylates, fatty acid ethoxylates, monoalkaolamide ethoxylates, sorbitan ester ethoxylates, fatty amine ethoxylates and ethylene oxide-propylene oxide copolymers (sometimes referred to as polymeric surfactants). Another important class of nonionics are the multihydroxy products such as glycol esters, glycerol (and polyglycerol) esters, glucosides (and polyglucosides) and sucrose esters. Amine oxides and sulfinyl surfactants represent nonionics with a small head group.

As mentioned above, surfactants are used for the formulation of all agrochemicals. For example, emulsifiable concentrates (EC's) are produced by mixing an agrochemical oil with another one such as xylene or trimethylbenzene or a mixture of various hydrocarbon solvents. Alternatively, a solid pesticide could be dissolved in a specific oil to produce a concentrated solution. In some cases, the pesticide oil may be used without any extra addition of oils. In all cases, a surfactant system (usually a mixture of two or three components) is added for a number of purposes. Firstly, the surfactant enables self-emulsification of the oil on addition to water. This occurs by a complex mechanism that involves a number of physical changes such as lowering of the interfacial tension at the oil/water interface, enhancement of turbulence at that interface with the result of spontaneous production of droplets. Secondly, the surfactant film that adsorbs at the oil/water interface stabilizes the produced emulsion against flocculation and/or coalescence. Emulsion breakdown must be prevented, otherwise excessive creaming or sedimentation or oil separation may take place during application. This results in an inhomogeneous application of the agrochemical on the one hand, and possible losses on the other. In recent years, there has been great demand to replace EC's with concentrated aqueous oil-in-water (o/w) emulsions, technically referred to as EW's. Several advantages may be envisaged for such replacements. In the first place, one is able to replace the added oil with water, which is of course much cheaper and environmentally acceptable. Secondly, removal of the oil could help in reducing undesirable effects such as phytotoxicity, skin irritation, etc. Thirdly, by formulating the pesticide as an o/w emulsion, one is able to control the droplet size to an optimum value which may be crucial for biological efficacy. Fourthly, water-soluble surfactants, which may be desirable for biological optimization, can be added to the aqueous continuous phase. The choice of a surfactant or a mixed surfactant system is crucial for preparation of a stable o/w emulsion. In recent years, macromolecular surfactants have been designed to produce very stable o/w emulsions which could be easily diluted into water and applied without any detrimental effects to the emulsion droplets.

Another agrochemical formulation where surfactants are used is the suspension concentrate (SC). Indeed, SC's are probably the most widely used systems in agrochemical formulations. Again, SC's are much more convenient to apply than wettable powders (WP's). Dust hazards are absent, and the formulation can be simply diluted in the spray tanks, without the need of any vigorous agitation. SC's are produced by a two- or three-stage process. The agrochemical powder is first dispersed in an aqueous solution of a surfactant or a macromolecule (usually referred to as the dispersing agent) using a high speed mixer. The surfactant used must ensure complete wetting of the powder by the aqueous medium. Both external and internal surfaces of the powder aggregates or agglomerates must be completely wetted to ensure complete dispersion of the powder into single particles. The resulting suspension is then subjected to a wet milling process (usually bead milling) to break any remaining aggregates or agglomerates and reduce the particle size to smaller values. One usually aims at a particle size distribution ranging from 0.1 to 5 μm, with an average of 1–2 μm. The surfactant or polymer added adsorbs on the particle surfaces, resulting in their colloidal stability. The particles need to be maintained stable over a long period of time, since any strong aggregation in the system may cause various problems. Firstly, the aggregates being larger than the primary particles tend to settle faster. Secondly, any gross aggregation may result in lack of dispersion on dilution. The large aggregates can block the spray nozzles and may reduce biological efficacy as a result of the inhomogeneous distribution of the particles on the target surface. Apart from their role in ensuring the colloidal stability of the suspension, surfactants are added to many SC's to enhance their biological efficacy. This is usually produced by solubilization of the insoluble compound in the surfactant micelles. This will be discussed in later sections. Another role a surfactant may play in SC's, is the reduction of crystal growth (Ostwald ripening). The latter process may occur when the solubility of the agrochemical is appreciable (say greater than 100 ppm) and when the SC is polydisperse. The smaller particles will have higher solubility than the larger ones. With time, the small particles dissolve and become deposited on the larger ones. Surfactants may reduce this Ostwald ripening by adsorption on the crystal surfaces, thus preventing deposition of the molecules at the surface.

Mixtures of suspensions and emulsions that are referred to as suspoemulsions have been formulated to allow application of two active ingredients: one being solid and the other is an immiscible liquid. Such multiphase systems are difficult to formulate due to the complex interaction between the suspension particles and emulsion droplets. These complex formulations require adequate selection of surfactants and emulsifiers to prevent any homoflocculation of the suspension particles, coalescence of the emulsion droplets and heteroflocculation between the particles and oil droplets.

Very recently, microemulsions have started to be considered as potential systems for formulating agrochemicals. Microemulsions are isotropic, thermodynamically stable systems consisting of oil, water and surfactant(s) whereby the free energy of formation of the system is zero or negative. It is obvious why such systems, if they can be formulated, are very attractive since they will have an indefinite shelf life (within a certain temperature range). Since the droplet size of microemulsions is very small (usually less than 50 nm), they appear transparent. The microemulsion droplets may be considered as swollen micelles and hence they will solubilize the agrochemical. This may result in considerable enhancement of the biological efficacy. Thus, microemulsions may offer several advantages over the commonly used macroemulsions. Unfortunately, formulating the agrochemical as a microemulsion is not straightforward since one usually uses two or more surfactants, an oil and the agrochemical. These tertiary systems produce various complex phases and it is essential to investigate the phase diagram before arriving at the optimum composition of microemulsion formation. A high concentration of surfactant (10–20 %) is needed to produce such a formulation. This makes such systems relatively more expensive to produce when compared to macroemulsions. However, the extra cost incurred could be offset by an enhancement of biological efficacy which means that a lower agrochemical application rate could be achieved.

Another important application of surfactants is that of controlled release formulations that is obtained by encapsulation. There are generally two mechanisms for release of the active ingredient (a.i.) from a capsule:
1. diffusion of the a.i. through the microcapsule wall;
2. destruction of the microcapsule wall by either physical means, e.g. mechanical power, or by chemical means, e.g. hydrolysis, biodegradation, thermal degradation, etc.

The release behavior is controlled by several factors such as particle size, wall thickness, type of wall material, wall structure (porosity, degree of polymerization, crosslink density, additives, etc.), type of core material (chemical structure, physical state, presence or absence of solvents) and amount or concentration of the core material. The release behavior is determined by interaction of these factors and optimization is essential for achieving the desirable release rate.

Controlled release formulation of agrochemicals offers a number of advantages of which the following are worth mentioning:

1. improvement of residual activity;
2. reduction of application dosage;
3. stabilization of the core active ingredient (a.i.) against environmental degradation;
4. reduction of mammalian toxicity by reducing worker exposure;
5. reduction of phytotoxicity;
6. reduction of fish toxicity;
7. reduction of environmental pollution.

One of the main advantages of using controlled release formulations, in particular microcapsules, is the reduction of physical incompatibility when mixtures are used in the spray tank. They also can reduce biological antagonism when mixtures are applied in the field.

Microencapsulation of agrochemicals is mainly carried out by interfacial condensation, in situ polymerization and coacervation. Interfacial condensation is perhaps the most widely used method for encapsulation in industry. The a.i., which may be oil soluble, oil dispersible or an oil itself, is first emulsified in water using a convenient surfactant or polymer. A hydrophobic monomer A is placed in the oil phase (oil droplets of the emulsion) and a hydrophilic monomer B is placed in the aqueous phase. The two monomers interact at the interface between the oil and the aqueous phase forming a capsule wall around the oil droplet. Two main types of systems may be identified. For example, if the material to be encapsulated is oil soluble, oil-dispersible or an oil itself, an oil-in-water (O/W) emulsion is first prepared. In this case the hydrophobic monomer is dissolved in the oil phase which forms the dispersed phase. The role of surfactant in this process is crucial since an oil-water emulsifier (with high hydrophilic-lipophilic balance, HLB) is required. Alternatively, a polymeric surfactant such as partially hydrolyzed polyvinyl acetate (referred to as polyvinyl alcohol, PVA) or an ethylene oxide-propylene oxide-ethylene oxide, PEO–PPO–PEO (Pluronic) block copolymer can be used. The emulsifier controls the droplet size distribution and hence the size of capsules formed. On the other hand, if the material to be encapsulated is water soluble, a water-in-oil (W/O) emulsion is prepared using a surfactant with low HLB number or an A–B–A block copolymer of polyhydroxystearic acid-polyethylene oxide-polyhydroxystearic acid (PHS-PEO-PHS). In this case the hydrophilic monomer is dissolved in the aqueous internal phase droplets.

In interfacial polymerization, the monomers A and B are polyfunctional monomers capable of causing polycondensation or polyaddition reaction at the interface. Examples of oil-soluble monomers are polybasic acid chloride, bis-haloformate and polyisocyantates, whereas water-soluble monomers can be polyamine or polyols.

Thus, a capsule wall of polyamide, polyurethane or polyurea may be formed. Some trifunctional monomers are present to allow crosslinking reactions. If water is the second reactant with polyisocyanates in the organic phase, polyurea walls are formed. The latter modification has been termed in situ interfacial polymerization.

One of the most useful microencapsulation processes involves reactions that produce formation of urea-formaldehyde (UF) resins. Urea, along with other ingredients such as amines, maleic anhydride copolymers or phenols, is added to the aqueous phase that contains oily droplets of the active ingredient that is to be encapsulated. Formaldehyde or formaldehyde oligomers are added and the reaction conditions are adjusted to form UF condensates, sometimes referred to as aminoplasts, that should preferentially wet the disperse phase. The reaction is continued to completion over several hours. Fairly high activity products can be obtained. A modification of this technique is the use of ethirified UF resins. The UF prepolymers are dissolved in the organic phase, along with the active ingredient, through the use of protective colloids (such as PVA), and the reaction is initiated through temperature and acid catalyst. This promotes the formation of the shell in the organic phase adjacent to the interface between the bulk-oil phase droplets and the aqueous phase solution.

It should be mentioned that the role of surfactants in the encapsulation process is very important. Apart from their direct role in the preparation of microcapsules dispersions, surfactants can be used to control the release of the active ingredient (a.i) from the microcapsule dispersion.

The third role of the surfactant system in agrochemicals is in enhancement of biological efficacy. It is essential to arrive at optimum conditions for effective use of the agrochemicals.

Optimization of the transfer of the agrochemical to the target requires careful analysis of the steps involved during application. Most agrochemicals are applied as liquid sprays, particularly for foliar application. The spray volume applied ranges from high values of the order of 1,000 liters per hectare (whereby the agrochemical concentrate is diluted with water) to ultralow volumes of the order of 1 liter per hectare (when the agrochemical formulation is applied without dilution). Various spray application techniques are used, of which spraying using hydraulic nozzles is probably the most common. In this case, the agrochemical is applied as spray droplets with a wide spectrum of droplet sizes (usually in the range 100–400 μm in diameter). On application, parameters such as droplet size spectrum, their impaction and adhesion, sliding and retention, wetting and spreading are of prime importance in ensuring maximum capture by the target surface as well as adequate coverage of the target surface. In addition to these "surface chemical" factors, i.e. the interaction with various interfaces, other parameters that affect biological efficacy are deposit formation, penetration and interaction with the site of action. Deposit formation, i.e. the residue left after evaporation of the spray droplets, has a direct effect on the efficacy of the pesticide, since such residues act as "reservoirs" of the agrochemical and hence they control the efficacy of the chemical after application. The penetration of the agrochemical and its interaction with the site of action is very important for systemic compounds. Enhancement of penetration is sometimes crucial to avoid removal of the agrochemical by environmental conditions such as rain and or/wind. All these factors are influenced by surfactants and polymers. In addition, some adjuvants that are used in combination with the formulation consist of oils and/or surfactant mixtures. Both statics and dynamics, e.g. static and dynamic surface tension and contact angles, as well as their effect on penetration and uptake of the chemical need to be investigated.

There are generally two main approaches for selection of adjuvants:

1. An interfacial (surface) physicochemical approach which is designed to increase the dose of the agrochemical received by the target plant or insect, i.e. enhancement of spray deposition, wetting, spreading, adhesion and retention.
2. Uptake activation that is enhanced by addition of surfactant which is the result of specific interactions between the surfactant, the agrochemical and the target species. These interactions may not be related to the intrinsic surface active properties of the surfactant/adjuvant.

The above two approaches must be considered when selecting an adjuvant for a given agrochemical and the type of formulation that is being used. The most important adjuvants are:

1. surface active agents;
2. polymers.

In some cases these are used in combination with crop oils (e.g. methyl oleate). Several complex recipes may be used and in many cases the exact composition of an adjuvant is not exactly known.

Adjuvants are applied in two different ways:

1. being incorporated in the formulation, mostly the case with flowables (SC's and EW's);
2. used in tank mixtures during application. Such adjuvants can be complex mixtures of several surfactants, oils, polymers, etc.

The choice of an adjuvant depends on:

1. the nature of the agrochemical: water soluble or insoluble (lipophilic) whereby its solubility and log P values are important;
2. the mode of action of the agrochemical, i.e. systemic or non-systemic, selective or non-selective;
3. the type of formulation that is used, i.e. flowable, EC, grain, granule, capsule, etc.

The most important adjuvants are surface active agents of the anionic, nonionic or zwitterionic type. In some cases polymers are added as stickers or anti-drift agents. The surfactant molecules accumulate at various interfaces as a result of their dual nature. This results in lowering of the air/liquid surface tension, γ_{LV}, and the solid/liquid interfacial tension, γ_{SL}. As the surfactant concentration is gradually increased, both γ_{LV} and γ_{SL} decrease until the critical micelle concentration (cmc) is reached, after which both values remain virtually constant. This situation represents the conditions under equilibrium whereby the rate of adsorption and desorption are the same. The situation under dynamic conditions, such as during spraying may be more complicated since the rate of adsorption is not equal to the rate of formation of droplets. Above the cmc, micelles are produced, which at low C values are essentially spherical (with an aggregation number in the region of 50–100 monomers). Depending on the conditions (e.g. temperature, salt concentration, structure of the surfactant molecules) other shapes may be produced, e.g. rod-shaped and lamellar micelles. Since micelles play a vital role when considering adjuvants, it is essential to understand their properties in some detail. Micelle formation is a dynamic process, i.e. a dynamic equilibrium is set up whereby surface active agent molecules are constantly leaving the micelles while others enter the micelles (same applies to the counterions). The dynamic process of micellization is described by two relaxation processes:

1. a short relaxation time τ_1 (of the order of 10^{-8}–10^{-3} s), which is the lifetime for a surfactant molecule in a micelle;
2. a longer relaxation time τ_2 (of the order of 10^{-3}–1 s), which is a measure of the micellization-dissolution process.

τ_1 and τ_2 depend on the surfactant structure, its chain length and these relaxation times determine some of the important factors in selecting adjuvants, such as the dynamic surface tension.

The cmc of nonionic surfactants is usually two orders of magnitude lower than the corresponding anionic of the same alkyl chain length. This explains why nonionics are generally preferred when selecting adjuvants. For a given series of nonionics, with the same alkyl chain length, the cmc decreases with decreasing the number of ethylene oxide (EO) units in the chain. Under equilibrium, the γ – log C curves shift to lower values as the EO chain length decreases. However, under dynamic conditions, the situation may be reversed, i.e. the dynamic surface tensions could become lower for the surfactant with the longer EO chain. This trend is understandable if one considers the dynamics of micelle formation. The surfactant with the longer EO chain has a higher cmc and it forms smaller micelles when compared with the surfactant containing the shorter EO chain. This means that the lifetime of a micelle with a longer EO chain is shorter than that with a longer EO chain. This explains why the dynamic surface tension of a solution of a surfactant containing a longer EO chain can be lower than that of a solution of an analogous surfactant (at the same concentration) with a shorter EO chain.

For a series of anionic surfactants with the same ionic head group, the lifetime of a micelle decreases with decreasing the alkyl chain length of the hydrophobic component. Branching of the alkyl chain could also play an important role in the lifetime of a micelle. It is, therefore, important to carry out dynamic surface tension measurements when selecting a surfactant as an adjuvant as this may play an important role in spray retention.

However, the above measurements should not be taken in isolation as other factors may also play an important role, e.g. solubilization which may require larger micelles. The selection of a surfactant as an adjuvant requires knowledge of the factors involved which will be discussed in some detail below.

At high surfactant concentrations (usually above 10 %) several liquid crystalline phases are produced. Three main types of liquid crystals may be distinguished:

1. hexagonal (middle) phase that consists of cylindrical anisotropic units with high viscosity;
2. cubic, body-centered isotropic phase with a viscosity that is higher than the hexagonal phase;
3. lamellar (neat) phase consisting of sheet-like units that are anisotropic, but with a viscosity that is lower than the hexagonal phase.

The above phases may form during evaporation of a spray drop. In some cases a middle phase is first produced which on further evaporation may produce a cubic phase which in view of its very high viscosity may entrap the agrochemical. This could be advantageous for some of the systemic fungicides which require "deposits" that act as reservoirs for the chemical. The viscous cubic phases may also enhance the tenacity of the agrochemical particles (particularly with SC's) and hence enhance rain fastness. In some other applications, a lamellar phase is preferred as this provides some mobility (due to its lower viscosity).

The application of an agrochemical, as a spray, involves a number of interfaces, where the interaction with the formulation plays a vital role. The first interface during application is that between the spray solution and the atmosphere (air) which governs the droplet spectrum, rate of evaporation, drift, etc. In this respect, the rate of adsorption of the surfactant and/or polymer at the air/liquid interface is of vital importance. This requires dynamic measurements of parameters such as surface tension which will give information on the rate of adsorption. The second interface is that between the impinging droplets and the leaf surface (with insecticides the interaction with the insect surface may be important). The droplets impinging on the surface undergo a number of processes that determine their adhesion and retention and further spreading on the target surface. The rate of evaporation of the droplet and the concentration gradient of the surfactant across the droplet govern the nature of the deposit formed. These processes of impaction, adhesion, retention, wetting and spreading and interaction with the leaf surface are all affected by the nature and concentration of the surfactant used.

11.4 Surfactants in paints and coatings [6]

Paints or surface coatings are complex multiphase colloidal systems that are applied as a continuous layer to a surface. A paint usually contains pigmented materials to distinguish it from clear films that are described as lacquers or varnishes. The main purpose of a paint or surface coating is to provide aesthetic appeal as well as to protect the surface. For example, a motor car paint can enhance the appearance of the car body by providing color and gloss and it also protects the car body from corrosion.

When considering a paint formulation one must know the specific interaction between the paint components and substrates. This subject is of particular importance when one considers the deposition and adhesion of the components to the substrate. The latter can be wood, plastic, metal, glass, etc. The interaction forces between the paint components and the substrate must be considered when formulating any paint. In addition the method of application can vary from one substrate and another. All these factors are affected by surfactants.

To obtain the fundamental understanding of the basic concepts and the role of surfactants, one must consider first the paint components. Most paint formulations consist of disperse systems (solid in liquid dispersions). The disperse phase consists of primary pigment particles (organic or inorganic) which provide the opacity, color and other optical effects. These are usually in the submicron range. Other coarse particles (mostly inorganic) are used in the primer and undercoat to seal the substrate and enhance adhesion of the top coat. The continuous phase consist of a solution of polymer or resin which provides the basis of a continuous film that seals the surface and protects it from the outside environment. Most modern paints contain latexes which are used as film formers. These latexes (with a glass transition temperature mostly below ambient temperature) coalesce on the surface and form a strong and durable film. Other components may be present in the paint formulation such as corrosion inhibitors, driers, fungicides, etc.

The primary pigment particles (normally in the submicron range) are responsible for the opacity, color and anti-corrosive properties. The principal pigment in use is titanium dioxide due to its high refractive index and the one that is used to produce white paint. To produce maximum scattering the particle size distribution of titanium dioxide has to be controlled within a narrow limit. Rutile with a refractive index of 2.76 is preferred over anatase that has a lower refractive index of 2.55. Thus, rutile gives the possibility of higher opacity than anatase and it is more resistant to chalking on exterior exposure. To obtain maximum opacity the particle size of rutile should be within 220–140 nm. The surface of rutile is photoactive and it is surface coated with silica and alumina in various proportions to reduce its photoactivity.

Colored pigments may consist of inorganic or organic particles. For a black pigment one can use carbon black, copper carbonate, manganese dioxide (inorganic) or aniline black (organic). For yellow one can use lead, zinc, chromates, cadmium sulfide, iron oxides (inorganic) or nickel azo yellow (organic). For blue/violet one can use ultramarine, Prussian blue, cobalt blue (inorganic) or phthalocyanin, indanthrone blue, carbazol violet (organic). For red one can use red iron oxide, cadmium selenide, red lead, chrome red (inorganic) or toluidine red, quinacridones (organic).

The dispersion of the pigment powder in the continuous medium requires several processes, namely wetting of the external and internal surface of the aggregates and agglomerates, separation of the particles from these aggregates and agglomerates by application of mechanical energy, displacement of occluded air and coating of the particles with the dispersion resin. It is also necessary to stabilize the particles against flocculation either by electrostatic double layer repulsion and/or steric repulsion. All these processes require the use of a surfactant.

The dispersion medium can be aqueous or nonaqueous depending on application. It consists of a dispersion of the binder in the liquid (which is sometimes referred to as the diluent). The term solvent is frequently used to include liquids that do not dissolve the polymeric binder. Solvents are used in paints to enable the paint to be made and they enable application of the paint to the surface. In most cases the solvent is removed after application by simple evaporation and if the solvent is completely removed from the paint film it should not affect the paint film performance. However, in the early life of the film solvent retention it can affect hardness, flexibility and other film properties. In water-based paints, the water may act as a true solvent for some of the components but it should be a non-solvent for the film former. This is particularly the case with emulsion paints.

With the exception of water, all solvents, diluents and thinners used in surface coatings are organic liquids with low molecular weight. Two types can be distinguished, hydrocarbons (both aliphatic and aromatic) and oxygenated compounds such as ethers, ketones, esters, ether alcohols, etc. Solvents, thinners and diluents control the flow of the wet paint on the substrate to achieve a satisfactory smooth, even thin film, which dries in a predetermined time. In most cases, mixtures of solvents are used to obtain the optimum conditions for paint application. The main factors that must be considered when choosing solvent mixtures are their solvency, viscosity, boiling point, evaporation rate, flash point, chemical nature, odor and toxicity.

As mentioned above, the dispersion medium consists of a solvent or diluent and the film former. The latter is also sometimes referred to as a "binder", since it functions by binding the particulate components together and this provides the continuous film-forming portion of the coating. The film former can be a low molecular weight polymer (oleoresinous binder, alkyd, polyurethane, amino resins, epoxide resin, unsaturated polyester), a high molecular weight polymer (nitrocellulose, solution vinyls, solution acrylics), an aqueous latex dispersion (polyvinyl acetate, acrylic or styrene/butadiene) or a nonaqueous polymer dispersion (NAD). The polymer solution may exist in the form of a fine particle dispersion in non-solvent. In some cases the system may be mixed solution/dispersion implying that the solution contains both single polymer chains and aggregates of these chains (sometimes referred to as micelles). A striking difference between a polymer that is completely soluble in the medium and that which contains aggregates of that polymer is the viscosity reached in both cases. A polymer that is completely soluble in the medium will show a higher viscosity at a given concentration compared to another polymer (at the same concentration) that produces aggregates. Another important difference is the rapid increase in the solution viscosity with increasing molecular weight for a completely soluble polymer. If the polymer makes aggregates in solution, increasing molecular weight of the polymer does not show a dramatic increase in viscosity.

The earliest film-forming polymers used in paints were based on natural oils, gums and resins. Modified natural products are based on cellulose derivatives such as nitrocellulose which is obtained by nitration of cellulose under carefully specified conditions. Organic esters of cellulose such as acetate and butyrate can also be produced. Another class of naturally occurring film formers are those based on vegetable oils and their derived fatty acids (renewable resource materials). Oils used in coatings include linseed oil, soya bean oil, coconut oil and tall oil. When chemically combined into resins, the oil contributes flexibility and with many oils oxidative crosslinking potential. The oil can also be chemically modified, for example the hydrogenation of castor oil can be combined with alkyd resins to produce some specific properties of the coating.

Another early binder used in paints are the oleoresinous vehicles that are produced by heating together oils and either natural or certain preformed resins, so that the resin dissolves or disperses in the oil portion of the vehicle. However these oleoresinous vehicles were later replaced by alkyd resins which are probably one of the first applications of synthetic polymers in the coating industry. These alkyd resins are polyesters obtained by reaction of vegetable oil triglycerides, polyols (e.g. glycerol) and dibasic acids or their anhydrides. These alkyd resins enhanced the mechanical strength, drying speed and durability over and above those obtained using the oleoresinous vehicles. The alkyds were also modified by replacing part of the dibasic acid with a diisocyanate (such as toluene diisocyanate, TDI) to produce greater toughness and quicker drying characteristics.

Another type of binder is based on polyester resins (both saturated and unsaturated). These are typically composed mainly of co-reacted di- or polyhydric alcohols and di- or tri-basic acid or acid anhydride. They have also been modified using silicone to enhance their durability.

More recently, acrylic polymers have been used in paints due to their excellent properties of clarity, strength and chemical and weather resistance. Acrylic polymers refer to systems containing acrylate and methylacrylate esters in their structure along with other vinyl unsaturated compounds. Both thermoplastic and thermosetting systems can be made, the latter are formulated to include monomers possessing additional functional groups that can further react to give crosslinks following the formation of the initial polymer structure. These acrylic polymers are synthesized by radical polymerization. The main polymer-forming reaction is a chain propagation step which follows an initial initiation process. A variety of chain transfer reactions are possible before chain growth ceases by a termination process.

Radicals produced by transfer, if sufficiently active, can initiate new polymer chains where a monomer is present which is readily polymerized. Radicals produced by chain transfer agents (low molecular weight mercaptans, e.g. primary octyl mercaptan) are designed to initiate new polymer chains. These agents are introduced to control the molecular weight of the polymer.

The monomers used for preparation of acrylic polymers vary in nature and can generally be classified as "hard" (such as methylmethacrylate, styrene and vinyl acetate) or "soft" (such as ethyl acrylate, butyl acrylate, 2-ethyl hexyl acrylate). Reactive monomers may also have hydroxyl groups (such as hydroxy ethyl acrylate). Acidic monomers such as methacrylic acid are also reactive and may be included in small amounts in order that the acid groups may enhance pigment dispersion. The practical coating systems are usually copolymers of "hard" and "soft". The polymer hardness is characterized by its glass transition temperature, T_g.

The vast majority of acrylic polymers consist of random copolymers. By controlling the proportion of "hard" and "soft" monomers and the molecular weight of the final copolymer one arrives at the right property that is required for a given coating. Two types of acrylic resins can be produced, namely thermoplastic and thermosetting. The former find application in automotive topcoats although they suffer from some disadvantages like cracking in cold conditions and this may require a process of plasticization. These problems are overcome by using thermosetting acrylics which improve the chemical and alkali resistance. Also it allows one to use higher solid contents in cheaper solvents. Thermosetting resins can be self-crosslinking or may require a co-reacting polymer or hardener.

Emulsion polymers (latexes) are the most commonly used film formers in the coating industry. This is particularly the case with aqueous emulsion paints that are used for home decoration. These aqueous emulsion paints are applied at room temperature and the latexes coalesce on the substrate forming a thermoplastic film. Sometimes functional polymers are used for crosslinking in the coating system. The polymer particles are typically submicron (0.1–0.5 μm).

Generally speaking, there are three methods for preparation of polymer dispersions, namely emulsion, dispersion and suspension polymerization. In emulsion polymerization, monomer is emulsified in a non-solvent, commonly water, usually in the presence of a surfactant. A water-soluble initiator is added, and particles of polymer form and grow in the aqueous medium as the reservoir of the monomer in the emulsified droplets is gradually used up. In dispersion polymerization (which is usually applied for preparation of nonaqueous polymer dispersion, commonly referred to as nonaqueous dispersion polymerization, NAD) monomer, initiator, stabilizer (referred to as protective agent) and solvent initially form a homogeneous solution. The polymer particles precipitate when the solubility limit of the polymer is exceeded. The particles continue to grow until the monomer is consumed. In suspension polymerization the monomer is emulsified in the continuous phase using a surfactant or polymeric suspending agent. The initiator (which is oil soluble) is dissolved in the monomer droplets and the droplets are converted into insoluble particles, but no new particles are formed.

As mentioned above, in emulsion polymerization, the monomer, e.g. styrene or methyl methacrylate that is insoluble in the continuous phase, is emulsified using a surfactant that adsorbs at the monomer/water interface. The surfactant micelles in bulk solution solubilize some of the monomer. A water-soluble initiator such as potassium persulfate $K_2S_2O_8$ is added and this decomposes in the aqueous phase forming free radicals that interact with the monomers forming oligomeric chains. It has long been assumed that nucleation occurs in the "monomer swollen micelles". The reasoning behind this mechanism was the sharp increase in the rate of reaction above the critical micelle concentration and that the number of particles formed and their size depend to a large extent on the nature of the surfactant and its concentration (which determines the number of micelles formed). However, this mechanism was later disputed and it was suggested that the presence of micelles means that excess surfactant is available and molecules will readily diffuse to any interface.

The most accepted theory of emulsion polymerization is referred to as the coagulative nucleation theory; a two-step coagulative nucleation model has been proposed. In this process the oligomers grow by propagation and this is followed by a termination process in the continuous phase. A random coil is produced which is insoluble in the medium and this produces a precursor oligomer at the θ-point. The precursor particles subsequently grow primarily by coagulation to form true latex particles. Some growth may also occur by further polymerization. The colloidal instability of the precursor particles may arise from their small size, and the slow rate of polymerization can be due to reduced swelling of the particles by the hydrophilic monomer. The role of surfactants in these processes is crucial since they determine the stabilizing efficiency, and the effectiveness of the surface active agent ultimately determines the number of particles formed. This was confirmed by using surface active agents of different nature. The effectiveness of any surface active agent in stabilizing the particles was the dominant factor and the number of micelles formed was relatively unimportant.

A typical emulsion polymerization formulation contains water, 50 % monomer blended for the required T_g, surfactant (and often colloid), initiator, pH buffer and fungicide. Hard monomers with a high T_g used in emulsion polymerization may be vinyl acetate, methyl methacrylate and styrene. Soft monomers with a low T_g include butyl acrylate, 2-ethylhexyl acrylate, vinyl versatate and maleate esters. Most suitable monomers are those with low, but not too low, water solubility. Other monomers such as acrylic acid, methacrylic acid, adhesion-promoting monomers may be included in the formulation. It is important that the latex particles coalesce as the diluent evaporates. The minimum film forming temperature (MFFT) of the paint is a characteristic of the paint system. It is closely related to the T_g of the polymer but the latter can be affected by materials present such as surfactant and the inhomogeneity of the polymer composition at the surface. High T_g polymers will not coalesce at room temperature and in this case a plasticizer ("coalescing agent") such as benzyl alcohol is incorporated in the formulation to reduce the T_g of the polymer thus reducing the MFFT of the paint. Clearly, for any paint system one must determine the MFFT since, as mentioned above, the T_g of the polymer is greatly affected the ingredients in the paint formulation.

Several types of surfactants can be used in emulsion polymerization such as anionics (sulfates, sulfonates, phosphates), cationics (alkyl ammonium surfactants), zwitterionics (betaines) and nonionics (alcohol and alky phenol ethoxylates, sorbitan esters and their ethoxylates, amine oxides, alkyl glucosides).

The role of surfactants is two-fold, firstly to provide a locus for the monomer to polymerize and secondly to stabilize the polymer particles as they form. In addition, surfactants aggregate to form micelles (above the critical micelle concentration) and these can solubilize the monomers. In most cases a mixture of anionic and nonionic surfactant is used for optimum preparation of polymer latexes. Cationic surfactants are seldom used, except for some specific applications where a positive charge is required on the surface of the polymer particles.

In addition to surfactants, most latex preparations require the addition of a polymer (sometimes referred to as "protective colloid") such as partially hydrolyzed polyvinyl acetate (commercially referred to as polyvinyl alcohol, PVA), hydroxyethyl cellulose or a block copolymer of polyethylene oxide (PEO) and polypropylene oxide (PPO). These polymers can be supplied with various molecular weights or proportions of PEO and PPO.

11.5 Surfactants in detergents [7]

The main function of a detergent is to remove soil: water-soluble materials (inorganic salts, sugar, etc.), particles (metal oxides, carbonates, soot, etc.), fats and oils (animal fat, vegetable oil, sebum, grease, etc.), proteins (from blood, egg, milk, etc.), carbohydrates (starch), bleachable dyes (fruits, vegetables, coffee, tea, etc.). For any given washing technology, the detergency performance depends on specific interactions among substrate surface, soil and detergent components. Understanding these interactions allows one to develop efficient and economic detergents.

Physical removal of soil from a surface occurs as a result of nonspecific adsorption of surfactants on the various interfaces present and through specific adsorption of chelating agents on certain polar soil components. An indirect effect is caused by calcium ion exchange, whereby the release of calcium ions from soil deposits and fibers causes a loosening of the structure of the residue. Compression of the electrical double layers at interfaces is significant. All these effects work together to remove oily and particulate soils from textile substrate or solid surfaces.

Several interfacial properties must be considered in order to understand the mechanism of dirt removal: the air/water interface that determines the surface tension, wetting, foam generation, film elasticity; the liquid/liquid interface that determines the interfacial tension, interfacial viscosity, emulsification; the solid/liquid interface that determines the suspension stability after soil removal; the solid/solid interface that determines the adhesion of soil particles or oil droplets to the substrate.

Removal of dirt (liquid or solid) can be from "smooth" surfaces, e.g. in dishwashers or from porous or fibrous materials, e.g. from fabrics. A good cleaning agent or detergent must have three main functions:
1. good wetting power;
2. ability to remove the dirt into the bulk of the liquid or to assist this process;
3. ability to solubilize or disperse the dirt once removed and to prevent its redeposition on the clean surface.

To formulate a good detergent, one has to understand the various processes involved: wetting, removal of dirt, liquid soiling, prevention of redeposition of dirt.

The best wetting agents are not necessarily the best detergents. For best wetting one needs to lower the dynamic surface tension (which is the value at very short periods of time since the process occurs over very short time scales). This requires molecules with shorter chain alkyl chains (C_8) and surfactants with short relaxation times for the micelles (usually high HLB molecules).

For best detergency one requires molecules that give high surface activity (maximum lowering of the surface tension) and this requires molecules with $C_{12}-C_{14}$ chains. Higher alkyl chain surfactants are not desirable since they have high Krafft temperatures.

In practice, most detergents consist of a wide range of molecules with various alkyl chains length and different head groups (anionic or nonionic with a range of ethylene oxide units). A detergent formulation will also contain other ingredients such as foam inhibitors, builders to remove multivalent ions, polymers for prevention of redeposition, bleaching agents, enzymes, corrosion inhibitors, perfumes, colors, etc.

Dirt is generally oily in nature and contains particles of dust, soot and so on. Its removal requires replacement of the soil/surface interface (characterized by a tension γ_{SD}) with a solid/water interface (characterized by a tension γ_{SW}) and dirt/water interface (characterized by a tension γ_{DW}).

The work of adhesion between a particle of dirt and a solid surface, W_d, is given by

$$W_{SD} = \gamma_{DW} + \gamma_{SW} - \gamma_{SD} \quad 1$$

A schematic representation of dirt removal is given in Fig. 11.4.

The task of the detergent is to lower γ_{DW} and γ_{SW} which decreases γ_{SD} and facilitates the removal of dirt by mechanical agitation.

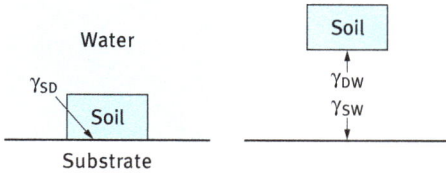

Fig. 11.4. Scheme of dirt removal.

Nonionic surfactants are generally less effective in removal of dirt than anionic surfactants. In practice a mixture of anionic and nonionic surfactants are used.

If the dirt is a liquid (oil or fat) its removal depends on the balance of contact angles. The oil or fat forms a low contact angle with the substrate (as illustrated in Fig. 11.5). To increase the contact angle between the oil and the substrate (with its subsequent removal), one has to reduce the substrate/water interfacial tension, γ_{SW} and oil/water interfacial tension γ_{DW}.

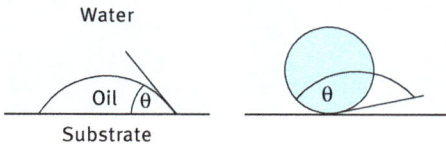

Fig. 11.5. Scheme of oil removal.

The addition of detergent increases the contact angle at the dirt/substrate/water interface so that the dirt "rolls up" and off the substrate. Surfactants that adsorb both at the substrate/water and the dirt/water interfaces are the most effective. If the surfactant adsorbs only at the dirt/water interface and lowers the interfacial tension between the oil and substrate (γ_{SD}) dirt removal is more difficult. Nonionic surfactants are the most effective in liquid dirt removal since they reduce the oil/water interfacial tension without reducing the oil/substrate tension.

To prevent dirt particles from redepositing on the substrate once they have been removed, they must be stabilized in the cleaning bath by colloid-chemical means. The prevention can be effected by electrical charge and/or steric barriers resulting from adsorption of the surfactant molecules from the cleaning bath both by the dirt particles and substrate. The most effective detergents for this purpose are nonionic surfactants of the poly(ethylene oxide) type. In some formulations, nonionic polymers or polyelectrolytes are added to prevent the redeposition of dirt particle (e.g. sodium carboxymthyl cellulose or other nonionic polymers).

References

[1] Th. F. Tadros, *Applied Surfactants*, Wiley-VCH, Germany, 2005.
[2] Th. F. Tadros (ed.), *Colloids in Cosmetics*, Wiley-VCH, Germany, 2008.
[3] D. Attwood and A. T. Florence, *Surfactant Systems*, Chapman and Hall, London, 1983.
[4] Th. F. Tadros, *Surfactants in Agrochemicals*, Marcel Dekker, New York, 1994.
[5] Th. F. Tadros, *Colloids in Agrochemicals*, Wiley-VCH, Germany, 2009.
[6] Th. F. Tadros, *Colloids in Paints*, Wiley-VCH, Germany, 2010.
[7] E. Smulders, *Laundry Detergents*, Wiley-VCH, Germany, 2002.

Index